Monte Carlo Simulation with Applications to Finance

CHAPMAN & HALL/CRC
Financial Mathematics Series

Aims and scope:
The field of financial mathematics forms an ever-expanding slice of the financial sector. This series aims to capture new developments and summarize what is known over the whole spectrum of this field. It will include a broad range of textbooks, reference works and handbooks that are meant to appeal to both academics and practitioners. The inclusion of numerical code and concrete real-world examples is highly encouraged.

Series Editors

M.A.H. Dempster
Centre for Financial Research
Department of Pure
Mathematics and Statistics
University of Cambridge

Dilip B. Madan
Robert H. Smith School
of Business
University of Maryland

Rama Cont
Center for Financial
Engineering
Columbia University
New York

Published Titles

American-Style Derivatives; Valuation and Computation, *Jerome Detemple*

Analysis, Geometry, and Modeling in Finance: Advanced Methods in Option Pricing,
  Pierre Henry-Labordère

An Introduction to Exotic Option Pricing, *Peter Buchen*

Credit Risk: Models, Derivatives, and Management, *Niklas Wagner*

Engineering BGM, *Alan Brace*

Financial Modelling with Jump Processes, *Rama Cont and Peter Tankov*

Interest Rate Modeling: Theory and Practice, *Lixin Wu*

Introduction to Credit Risk Modeling, Second Edition, *Christian Bluhm, Ludger Overbeck, and*
  Christoph Wagner

Introduction to Stochastic Calculus Applied to Finance, Second Edition,
  Damien Lamberton and Bernard Lapeyre

Monte Carlo Methods and Models in Finance and Insurance, *Ralf Korn, Elke Korn,*
  and Gerald Kroisandt

Monte Carlo Simulation with Applications to Finance, *Hui Wang*

Numerical Methods for Finance, *John A. D. Appleby, David C. Edelman, and John J. H. Miller*

Option Valuation: A First Course in Financial Mathematics, *Hugo D. Junghenn*

Portfolio Optimization and Performance Analysis, *Jean-Luc Prigent*

Quantitative Fund Management, *M. A. H. Dempster, Georg Pflug, and Gautam Mitra*

Risk Analysis in Finance and Insurance, Second Edition, *Alexander Melnikov*

Robust Libor Modelling and Pricing of Derivative Products, *John Schoenmakers*

Stochastic Finance: A Numeraire Approach, *Jan Vecer*

Stochastic Financial Models, *Douglas Kennedy*

Structured Credit Portfolio Analysis, Baskets & CDOs, *Christian Bluhm and Ludger Overbeck*

Understanding Risk: The Theory and Practice of Financial Risk Management, *David Murphy*

Unravelling the Credit Crunch, *David Murphy*

Proposals for the series should be submitted to one of the series editors above or directly to:
CRC Press, Taylor & Francis Group
4th, Floor, Albert House
1-4 Singer Street
London EC2A 4BQ
UK

Chapman & Hall/CRC FINANCIAL MATHEMATICS SERIES

Monte Carlo Simulation with Applications to Finance

Hui Wang

Brown University
Providence, Rhode Island, USA

CRC Press
Taylor & Francis Group
Boca Raton London New York

CRC Press is an imprint of the
Taylor & Francis Group, an **informa** business
A CHAPMAN & HALL BOOK

CRC Press
Taylor & Francis Group
6000 Broken Sound Parkway NW, Suite 300
Boca Raton, FL 33487-2742

First issued in paperback 2019

© 2012 by Taylor & Francis Group, LLC
CRC Press is an imprint of Taylor & Francis Group, an Informa business

No claim to original U.S. Government works

ISBN-13: 978-1-4398-5824-0 (hbk)
ISBN-13: 978-0-367-38135-6 (pbk)

Visit the Taylor & Francis Web site at
http://www.taylorandfrancis.com

and the CRC Press Web site at
http://www.crcpress.com

Preface

This book can serve as the text for a one-semester course on Monte Carlo simulation. The intended audience is advanced undergraduate students or students in master's programs who wish to learn the basics of this exciting topic and its applications to finance.

The book is largely self-contained. The only prerequisite is some experience with probability and statistics. Prior knowledge on option pricing is helpful but not essential. As in any study of Monte Carlo simulation, coding is an integral part and cannot be ignored. The book contains a large number of MATLAB® coding exercises. They are designed in a progressive manner so that no prior experience with MATLAB is required.

Much of the mathematics in the book is informal. For example, random variables are simply defined to be functions on the sample space, even though they should be measurable with respect to appropriate σ-algebras; exchanging the order of integrations is carried out liberally, even though it should be justified by the Tonelli–Fubini Theorem. The motivation for doing so is to avoid the technical measure theoretic jargon, which is of little concern in practice and does not help much to further the understanding of the topic.

The book is an extension of the lecture notes that I have developed for an undergraduate course on Monte Carlo simulation at Brown University. I would like to thank the students who have taken the course, as well as the Division of Applied Mathematics at Brown, for their support.

Hui Wang
Providence, Rhode Island
January, 2012

MATLAB® is a trademark of The MathWorks, Inc. For product information, please contact:

The MathWorks, Inc.
3 Apple Hill Drive
Natick, MA 01760-2098 USA
Tel: 508 647 7000
Fax: 508-647-7001
E-mail: info@mathworks.com
Web: www.mathworks.com

Contents

Chapter 1

Review of Probability

Probability theory is the essential mathematical tool for the design and analysis of Monte Carlo simulation schemes. It is assumed that the reader is somewhat familiar with the elementary probability concepts such as random variables and multivariate probability distributions. However, for the sake of completeness, we use this chapter to collect a number of basic results from probability theory that will be used repeatedly in the rest of the book.

1.1 Probability Space

In probability theory, *sample space* is the collection of all possible outcomes. Throughout the book, the sample space will be denoted by Ω. A generic element of the sample space represents a possible outcome and is called a *sample point*. A subset of the sample space is called an *event*.

1. The empty set is denoted by $\emptyset$.

2. The complement of an event A is denoted by A^c. $A \cap B$

3. The intersection of events A and B is denoted by $A \cap B$ or simply AB.

4. The union of events A and B is denoted by $A \cup B$. $A \cup B$

A *probability measure* $\mathbb{P}$ on Ω is a mapping from the events of Ω to the real line $\mathbb{R}$ that satisfies the following three axioms:

(i) $\mathbb{P}(\Omega) = 1$.

Union $(A \cup B)$ Intersection $A \cap B$

(ii) $0 \leq \mathbb{P}(A) \leq 1$ for every event A.

(iii) For every sequence of *mutually exclusive* events $\{A_1, A_2, \ldots\}$, that is, $A_i \cap A_j = \emptyset$ for all $i \neq j$,

$$\mathbb{P}\left(\cup_{n=1}^\infty A_n\right) = \sum_{n=1}^\infty \mathbb{P}\left(A_n\right).$$

Lemma 1.1. *Let $\mathbb{P}$ be a probability measure. Then the following statements hold.*

1. $\mathbb{P}(A) + \mathbb{P}(A^c) = 1$ *for any event A.*

2. $\mathbb{P}(A \cup B) = \mathbb{P}(A) + \mathbb{P}(B) - \mathbb{P}(AB)$ *for any events A and B. More generally,*

$$\mathbb{P}(A_1 \cup \cdots \cup A_n) = \sum_i \mathbb{P}(A_i) - \sum_{i<j} \mathbb{P}(A_i A_j) + \sum_{i<j<k} \mathbb{P}(A_i A_j A_k)$$
$$+ \cdots + (-1)^{n+1} \mathbb{P}(A_1 \cdots A_n)$$

   for an arbitrary collection of events $A_1, \ldots, A_n$.

This lemma follows immediately from the three axioms. We leave the proof to the reader as an exercise.

1.2   Independence and Conditional Probability

Two events A and B are said to be *independent* if $\mathbb{P}(AB) = \mathbb{P}(A) \cdot \mathbb{P}(B)$. More generally, a collection of events $A_1, A_2, \cdots, A_n$ are said to be *independent* if

$$\mathbb{P}(A_{k_1} A_{k_2} \cdots A_{k_m}) = \mathbb{P}(A_{k_1}) \cdot \mathbb{P}(A_{k_2}) \cdot \cdots \cdot \mathbb{P}(A_{k_m})$$

for any $1 \leq k_1 < k_2 < \ldots < k_m \leq n$.

Lemma 1.2. *Suppose that events A and B are independent. Then so are events A^c and B, A and B^c, and A^c and B^c. Similar results hold for an arbitrary collection of independent events.*

PROOF. Consider the events A^c and B. Since $(A^c B)$ and (AB) are disjoint and $(A^c B) \cup (AB) = B$, it follows that

$$\mathbb{P}(A^c B) + \mathbb{P}(AB) = \mathbb{P}(B).$$

By the independence of A and B, $\mathbb{P}(AB) = \mathbb{P}(A)\mathbb{P}(B)$. Therefore,

$$
\begin{aligned}
\mathbb{P}(A^cB) &= \mathbb{P}(B) - \mathbb{P}(AB) \\
&= \mathbb{P}(B) - \mathbb{P}(A)\mathbb{P}(B) \\
&= \mathbb{P}(B)[1 - \mathbb{P}(A)] \\
&= \mathbb{P}(B)\mathbb{P}(A^c).
\end{aligned}
$$

In other words, A^c and B are independent. The proof for other cases is similar and thus omitted. ∎

Consider two events A and B with $\mathbb{P}(B) > 0$. The *conditional probability* of A given B is defined to be

what's the chance of A occuring, given that B occurs.

$$
\longrightarrow \quad \mathbb{P}(A|B) = \frac{\mathbb{P}(AB)}{\mathbb{P}(B)}. \tag{1.1}
$$

When $\mathbb{P}(B) = 0$, the conditional probability $\mathbb{P}(A|B)$ is undefined. However, it is always true that

$$
\mathbb{P}(AB) = \mathbb{P}(A|B)\mathbb{P}(B),
$$

where the right-hand-side is defined to be 0 as long as $\mathbb{P}(B) = 0$.

Lemma 1.3. *Given any event B with $\mathbb{P}(B) > 0$, the following statements hold.*

1. *An event A is independent of B if and only if $\mathbb{P}(A|B) = \mathbb{P}(A)$.* So B has no bearing on whether A occurs or not.

2. *For any disjoint events A and C,* independent

$$
\mathbb{P}(A \cup C|B) = \mathbb{P}(A|B) + \mathbb{P}(C|B).
$$

3. *For any events A_1 and A_2,*

$$
\mathbb{P}(A_1A_2|B) = \mathbb{P}(A_1|B) \cdot \mathbb{P}(A_2|A_1B).
$$

PROOF. All these claims follow directly from the definition (1.1). We should only give the proof of (3). The right-hand-side equals

$$
\frac{\mathbb{P}(A_1B)}{\mathbb{P}(B)} \cdot \frac{\mathbb{P}(A_1A_2B)}{\mathbb{P}(A_1B)} = \frac{\mathbb{P}(A_1A_2B)}{\mathbb{P}(B)},
$$

which equals the left-hand-side. We complete the proof. ∎

Theorem 1.4. (Law of Total Probability). *Suppose that* $\{B_n\}$ *is a partition of the sample space, that is,* $\{B_n\}$ *are mutually exclusive and* $\cup_n B_n = \Omega$. *Then for any event A,*

$$P(A) = \sum_n P(A|B_n)P(B_n).$$

PROOF. Observe that $\{AB_n\}$ are disjoint events and $\cup_n AB_n = A$. It follows that

$$P(A) = \sum_n P(AB_n) = \sum_n P(A|B_n)P(B_n).$$

We complete the proof. ∎

Example 1.1. An investor has purchased bonds from five S&P AAA-rated banks and three S&P A-rated banks.

S&P rating	AAA	AA	A	BBB	BB	B	CCC
Probability	1	4	12	50	300	1100	2800

Annual default probability in basis points, 100 basis points = 1%

Assuming that all these banks are independent, what is the probability that

(a) at least one of the banks default?

(b) exactly one bank defaults?

SOLUTION: The probability that at least one of the banks default equals

$$
\begin{aligned}
1 - \mathbb{P}(\text{none of the banks default}) \ &= \ 1 - (1 - 0.0001)^5 \cdot (1 - 0.0012)^3 \\
&= \ 1 - 0.9999^5 \cdot 0.9988^3 \\
&= \ 40.94 \text{ bps.}
\end{aligned}
$$

Observe that there are five equally likely ways that exactly one AAA-rated bank defaults and three equally likely ways that exactly one A-rated bank defaults. Hence, the probability that exactly one bank defaults equals

$$5 \cdot 0.9999^4 \cdot 0.0001 \cdot 0.9988^3 + 3 \cdot 0.9999^5 \cdot 0.9988^2 \cdot 0.0012 = 40.88 \text{ bps.}$$

The answers to (a) and (b) are nearly identical because the probability that more than one bank will default is negligible. ∎

Example 1.2. A technical analyst has developed a simple model that uses the data from previous two days to predict the stock price movement of the following day. Let "+" and "−" denote the stock price movement in a trading day:

$$\text{"+"} \quad = \quad \text{stock price moves up or remains unchanged,}$$
$$\text{"−"} \quad = \quad \text{stock price moves down.}$$

Below is the probability distribution.

	Tomorrow	
(Yesterday, today)	+	−
(+, +)	0.2	0.8
(−, +)	0.4	0.6
(+, −)	0.7	0.3
(−, −)	0.5	0.5

Assume that the stock price movements yesterday and today are $(-, +)$. Compute the probability that the stock price movement will be "+" for

(a) tomorrow,

(b) the day after tomorrow.

SOLUTION: Define the following events:

$$A_1 \quad = \quad \text{stock price movement tomorrow is "+",}$$
$$A_2 \quad = \quad \text{stock price movement the day after tomorrow is "+",}$$
$$B \quad = \quad \text{stock price movements yesterday and today are } (-, +).$$

Then the probability that the stock price movement tomorrow will be "+" is

$$\mathbb{P}(A_1|B) = \mathbb{P}(+|-, +) = 0.4.$$

By Lemma 1.3, the probability that the stock price movement the day after tomorrow will be "+" is

$$
\begin{aligned}
\mathbb{P}(A_2|B) &= \mathbb{P}(A_1 A_2|B) + \mathbb{P}(A_1^c A_2|B) \\
&= \mathbb{P}(A_1|B) \cdot \mathbb{P}(A_2|A_1 B) + \mathbb{P}(A_1^c|B) \cdot \mathbb{P}(A_2|A_1^c B) \\
&= \mathbb{P}(+|-, +) \cdot \mathbb{P}(+|+, +) + \mathbb{P}(-|-, +) \cdot \mathbb{P}(+|+, -) \\
&= 0.4 \times 0.2 + 0.6 \times 0.7 \\
&= 0.5.
\end{aligned}
$$

1.3   Random Variables

A *random variable* is a mapping from the sample space to the real line $\mathbb{R}$. The *cumulative distribution function* (cdf) of a random variable X is defined by

$$F(x) = \mathbb{P}(X \leq x)$$

for every $x \in \mathbb{R}$. It is always nondecreasing and continuous from the right. Furthermore,

$$\lim_{x \to -\infty} F(x) = 0, \qquad \lim_{x \to +\infty} F(x) = 1.$$

1.3.1   Discrete Random Variables

A random variable is said to be *discrete* if it can assume at most countably many possible values. Suppose that $\{x_1, x_2, \cdots\}$ is the set of all possible values of a random variable X. The function

possible values

$$p(x_i) = \mathbb{P}(X = x_i), \quad i = 1, 2, \cdots$$

is called the *probability mass function* of X. The *expected value* (or *expectation*, *mean*) of X is defined to be

Probability mass function...

$$E[X] = \sum_i x_i p(x_i).$$

More generally, given any function $h : \mathbb{R} \to \mathbb{R}$, the expected value of the random variable $h(X)$ is given by

$$E[h(X)] = \sum_i h(x_i) p(x_i).$$

While the expected value measures the average of a random variable, the most common measure of the variability of a random variable is the *variance*, which is defined by

$$\text{Var}[X] = E[(X - E[X])^2] = E[X^2] - (E[X])^2.$$

The *standard deviation* of X is just the square root of the variance:

$$\text{Std}[X] = \sqrt{\text{Var}[X]}.$$

Among the most frequently used discrete random variables are Bernoulli random variables, binomial random variables, and Poisson random variables.

1. **Bernoulli with parameter p.** A random variable X that takes values in $\{0, 1\}$ and

These describe the distribution of many natural phenomena.

$$P(X = 1) = p, \quad P(X = 0) = 1 - p.$$

$$E[X] = p, \quad \mathrm{Var}[X] = p(1 - p).$$

2. **Binomial with parameters (n, p).** A random variable X that takes values in $\{0, 1, \ldots, n\}$ and

$$P(X = k) = \binom{n}{k} p^k (1 - p)^{n-k}.$$

$$E[X] = np, \quad \mathrm{Var}[X] = np(1 - p).$$

3. **Poisson with parameter λ.** A random variable X that takes values in $\{0, 1, \ldots\}$ and

$$P(X = k) = e^{-\lambda} \frac{\lambda^k}{k!}.$$

$$E[X] = \lambda, \quad \mathrm{Var}[X] = \lambda.$$

Example 1.3. Compare the following two scenarios. The default probability of an S&P B-rated bank is assumed to be 1100 basis points.

(a) A company invests 10 million dollars in the 1-year bonds issued by an S&P B-rated bank, with an annual interest rate of 10%. Compute the expectation and standard deviation of the value of these bonds at maturity.

(b) A company diversifies its portfolio by dividing 10 million dollars equally between the 1-year bonds issued by two different S&P B-rated banks. Assume that the two banks are independent and offer the same annual interest rate of 10%. Compute the expectation and standard variation of the value of these bonds at maturity.

SOLUTION: Let X (in millions) be the value of these bonds at maturity.

(a) It is easy to see that the distribution of X is given by the following table:

Value of X	11	0
Probability	0.89	0.11

Therefore,
$$E[X] = 11 \times 0.89 + 0 \times 0.11 = 9.79,$$

$$\text{Std}[X] = \sqrt{\text{Var}[X]} = \sqrt{0.89 \times 1.21^2 + 0.11 \times (-9.79)^2} = 3.442.$$

(b) If both banks default then $X = 0$. If one of the banks defaults then $X = 5.5$. If none of the banks default then $X = 11$. The respective probabilities are given by the following table:

Value of X	11	5.5	0
Probability	0.7921	0.1958	0.0121

For example, there are two equally likely ways that one of the banks defaults, and thus the probability of $X = 5.5$ is

$$2 \times 0.11 \times (1 - 0.11) = 0.1958$$

by independence. The expected value and standard deviation of X can be similarly calculated.

$$E[X] = 11 \times 0.7921 + 5.5 \times 0.1958 + 0 \times 0.0121 = 9.79,$$

$$\text{Std}[X] = \sqrt{\text{Var}[X]} = 2.434.$$

These two investment strategies have the same expected return. However, the standard deviation or variance of the second strategy is significantly smaller than that of the first strategy. If one defines variance as the *measure of risk*, then the second strategy has the same expected return but less risk. In other words, diversification reduces risk.   ∎

Example 1.4. Consider the following model for stock price. Denote by S_i the price at the i-th time step. If the current price is S, then at the next time step the price either moves up to uS with probability p or moves down to dS with probability $1 - p$. Here $d < 1 < u$ are given positive constants. Given $S_0 = x$, find the distribution of S_n.

SOLUTION: Suppose that among the first n time steps there are k steps at which the stock price moves up. Then there are $(n - k)$ steps at which the stock price moves down and

$$S_n = u^k d^{n-k} S_0 = u^k d^{n-k} x.$$

Since the number of time steps at which the stock price moves up is a binomial random variable with parameters n and p, the distribution of S_n is given by

$$\mathbb{P}(S_n = u^k d^{n-k} x) = \binom{n}{k} p^k (1-p)^{n-k},$$

for $k = 0, 1, \ldots, n$. ∎

1.3.2 Continuous Random Variables → can take on any value in the interval

A random variable X is said to be *continuous* if there exists a nonnegative function $f(x)$ such that

$$\mathbb{P}(X \in B) = \int_B f(x)\,dx$$

for any subset $B \subseteq \mathbb{R}$. The function f is said to be the *density* of X and must satisfy the equality

$$\int_{-\infty}^{\infty} f(x)\,dx = 1.$$

The relation between the cumulative distribution function F and the density f is given by

$$f(x) = F'(x), \quad F(x) = \int_{-\infty}^{x} f(t)\,dt.$$

The *expected value* (or *expectation, mean*) of X is defined to be

$$E[X] = \int_{-\infty}^{\infty} x f(x)\,dx.$$

More generally, for any function $h : \mathbb{R} \to \mathbb{R}$, the expected value of $h(X)$ is given by

$$E[h(X)] = \int_{-\infty}^{\infty} h(x) f(x)\,dx.$$

As in the discrete random variable case, the *variance* and *standard deviation* of X are defined as follows:

$$\mathrm{Var}[X] = E[(X - E[X])^2] = E[X^2] - (E[X])^2,$$

$$\mathrm{Std}[X] = \sqrt{\mathrm{Var}[X]}.$$

We should discuss some of the most widely used continuous random variables in finance: uniform random variables, exponential random variables, normal random variables, and lognormal random variables.

1. **Uniform** on $[a, b]$. A random variable X with density

$$f(x) = \begin{cases} (b-a)^{-1} & \text{if } x \in [a, b], \\ 0 & \text{otherwise.} \end{cases}$$

$cdf \left(\dfrac{x-a}{b-a} \right)$

$$E[X] = \frac{b+a}{2}, \quad \text{Var}[X] = \frac{(b-a)^2}{12}.$$

2. **Exponential with rate λ.** A random variable X with density

$$f(x) = \begin{cases} \lambda e^{-\lambda x} & \text{if } x \geq 0, \\ 0 & \text{otherwise.} \end{cases}$$

$$E[X] = \frac{1}{\lambda}, \quad \text{Var}[X] = \frac{1}{\lambda^2}.$$

3. **Normal with mean μ and variance σ^2 : $N(\mu, \sigma^2)$.** A random variable X with density

$$f(x) = \frac{1}{\sqrt{2\pi}\sigma} e^{-\frac{1}{2\sigma^2}(x-\mu)^2}, \quad x \in \mathbb{R}.$$

$$E[X] = \mu, \quad \text{Var}[X] = \sigma^2.$$

The special case where $\mu = 0$ and $\sigma = 1$ is referred to as the **standard normal**.

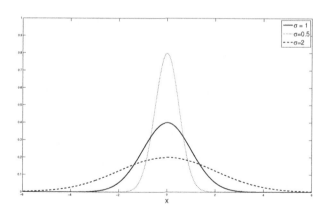

Figure 1.1: Density of $N(0, \sigma^2)$.

An important property of the normal distributions is that any linear transform of a normal random variable is still normal. More precisely, we have the following result, whose proof is left to Exercise 1.6.

Lemma 1.5. *Assume that X is $N(\mu, \sigma^2)$. Let a and b be two arbitrary constants. Then $a + bX$ is normal with mean $a + b\mu$ and variance $b^2\sigma^2$.*

An immediate corollary of the preceding lemma is that if X is $N(\mu, \sigma^2)$, then

$$Z = \frac{X - \mu}{\sigma}$$

is a standard normal random variable. Throughout the book, we will use Φ to denote the cumulative distribution function of the standard normal. That is,

calculating the area under the curve, the cumulative probability.

$$\Phi(x) = \mathbb{P}(N(0,1) \le x) = \int_{-\infty}^{x} \frac{1}{\sqrt{2\pi}} e^{-\frac{1}{2}z^2} \, dz.$$

Due to the symmetry of the standard normal density, for every $x \in \mathbb{R}$

$$\Phi(x) + \Phi(-x) = 1.$$

4. **Lognormal with parameters μ and σ^2 : LogN(μ, σ^2).** A positive random variable X whose natural logarithm is normally distributed with mean μ and variance σ^2. In other words,

$$X = e^Y, \quad Y = N(\mu, \sigma^2).$$

$$E[X] = e^{\mu + \frac{1}{2}\sigma^2}, \quad \mathrm{Var}[X] = \left(e^{\sigma^2} - 1\right) \cdot e^{2\mu + \sigma^2}.$$

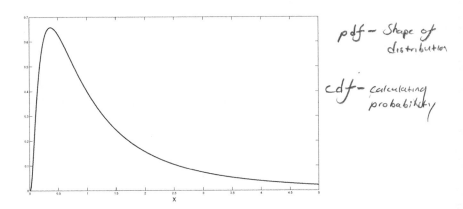

pdf – Shape of distribution

cdf – calculating probability

Figure 1.2: Density of LogN$(0, 1)$.

Example 1.5. Value-at-risk (VaR) measures, within a confidence level, the maximum loss a portfolio could suffer. To be more precise, denote by X the change in the market value of a portfolio during a given time period. Then for a given confidence level $1 - \alpha$ where $\alpha \in (0, 1)$, the VaR is defined by

$$\mathbb{P}(X \le -\text{VaR}) = \alpha.$$

In other words, with probability $1 - \alpha$, the maximum loss will not exceed VaR.

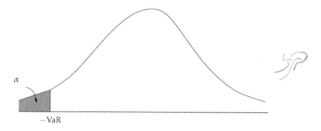

Figure 1.3: Value at Risk.

There is a simple formula for VaR when X is assumed to be normally distributed with mean μ and variance σ^2. Given a confidence level $1 - \alpha$,

$$\alpha = \mathbb{P}(X \le -\text{VaR}) = \mathbb{P}\left(\frac{X - \mu}{\sigma} \le \frac{-\text{VaR} - \mu}{\sigma}\right).$$

Since $(X - \mu)/\sigma$ is a standard normal random variable, it follows that

$$\frac{-\text{VaR} - \mu}{\sigma} = -z_\alpha,$$

where z_α is determined by $\Phi(-z_\alpha) = \alpha$. Therefore, $\text{VaR} = z_\alpha \sigma - \mu$. ∎

Example 1.6. The evaluation of call options often involves the calculation of expected values such as $E\left[(S - K)^+\right]$, where S is the price of the underlying stock, K is a given positive constant, and x^+ denotes the positive part of x, that is, $x^+ = \max\{x, 0\}$. Assuming that S is lognormally distributed with parameters μ and σ^2, compute this expected value.

SOLUTION: Since $X = \log S$ is distributed as $N(\mu, \sigma^2)$, it follows that

$$
\begin{aligned}
E[(S - K)^+] &= \int_{\log K}^{\infty} (e^x - K) \frac{1}{\sqrt{2\pi}\sigma} e^{-\frac{1}{2\sigma^2}(x-\mu)^2} \, dx \\
&= \int_{\theta}^{\infty} (e^{\mu + \sigma z} - K) \frac{1}{\sqrt{2\pi}} e^{-\frac{1}{2}z^2} \, dz,
\end{aligned}
$$

where in the second equality we have used the change of variable $x = \mu + \sigma z$ and let

$$\theta = \frac{\log K - \mu}{\sigma}.$$

Direct calculation yields that

$$
\begin{aligned}
\int_\theta^\infty e^{\mu+\sigma z} \frac{1}{\sqrt{2\pi}} e^{-\frac{1}{2}z^2} dz &= e^{\mu+\frac{1}{2}\sigma^2} \int_\theta^\infty \frac{1}{\sqrt{2\pi}} e^{-\frac{1}{2}(z-\sigma)^2} dz \\
&= e^{\mu+\frac{1}{2}\sigma^2} \int_{\theta-\sigma}^\infty \frac{1}{\sqrt{2\pi}} e^{-\frac{1}{2}z^2} dz \\
&= e^{\mu+\frac{1}{2}\sigma^2} \Phi(\sigma - \theta), \\
K \int_\theta^\infty \frac{1}{\sqrt{2\pi}} e^{-\frac{1}{2}z^2} dz &= K\Phi(-\theta).
\end{aligned}
$$

Hence,

$$E[(S-K)^+] = e^{\mu+\frac{1}{2}\sigma^2} \Phi(\sigma - \theta) - K\Phi(-\theta). \qquad \blacksquare$$

1.4 Random Vectors

A *random vector* is a collection of random variables defined on the same sample space. The components of a random vector can be discrete, continuous, or mixed. We will, for the moment, restrict ourselves to a two-dimensional random vector (X, Y). The extension to general random vectors is obvious.

1. **Discrete random vectors.** If both X and Y are discrete, it is convenient to define the *joint probability mass function* by

$$p(x_i, y_j) = \mathbb{P}(X = x_i, Y = y_j).$$

The probability mass functions for X and for Y are the same as the *marginal probability mass functions*

$$p_X(x_i) = \sum_j p(x_i, y_j), \quad p_Y(y_j) = \sum_i p(x_i, y_j),$$

respectively. For any function $h : \mathbb{R}^2 \to \mathbb{R}$, the expected value of $h(X, Y)$ is given by

$$E[h(X, Y)] = \sum_{i,j} h(x_i, y_j) p(x_i, y_j).$$

2. **Continuous random vectors.** X and Y are said to be *jointly continuous* if there exists a nonnegative *joint density function*, say $f(x, y)$, such that

$$\mathbb{P}(X \in A, Y \in B) = \iint_{A \times B} f(x, y)\, dx dy$$

for any $A, B \subseteq \mathbb{R}$. The density functions for X and for Y are the same as the *marginal density functions*

$$f_X(x) = \int_{-\infty}^{\infty} f(x, y)\, dy, \quad f_Y(y) = \int_{-\infty}^{\infty} f(x, y)\, dx,$$

respectively. For any function $h : \mathbb{R}^2 \to \mathbb{R}$, the expected value of $h(X, Y)$ is given by

$$E[h(X, Y)] = \iint_{\mathbb{R}^2} h(x, y) f(x, y)\, dx dy.$$

Let a and b be two arbitrary constants and consider the function $h(x, y) = ax + by$. We obtain the following result immediately.

Theorem 1.6. *For any random variables X and Y, any constants a and b,*

$$E[aX + bY] = aE[X] + bE[Y].$$

1.4.1   Covariance and Correlation

The *covariance* of random variables X and Y is defined to be

$$\mathrm{Cov}(X, Y) = E\left[(X - EX)(Y - EY)\right] = E[XY] - E[X]E[Y].$$

Direct calculation yields that for any random variables X, Y, Z and any constants a, b, c, the following relations hold:

$$
\begin{aligned}
\mathrm{Cov}(X, X) &= \mathrm{Var}[X], \\
\mathrm{Cov}(X, Y) &= \mathrm{Cov}(Y, X), \\
\mathrm{Cov}(a, X) &= 0, \\
\mathrm{Cov}(aX + bY, Z) &= a\mathrm{Cov}(X, Z) + b\mathrm{Cov}(Y, Z), \\
\mathrm{Cov}(X, aY + bZ) &= a\mathrm{Cov}(X, Y) + b\mathrm{Cov}(X, Z).
\end{aligned}
$$

We can now state the variance formula for sums of random variables. The proof is a straightforward application of the preceding identities and thus omitted.

Lemma 1.7. *For any random variables* $X_1, X_2, \ldots, X_n,$

$$\mathrm{Var}\left(\sum_{i=1}^{n} X_i\right) = \sum_{i=1}^{n} \mathrm{Var}[X_i] + \sum_{i \neq j} \mathrm{Cov}(X_i, X_j)$$

Variance for a sum of variables.

$$= \sum_{i=1}^{n} \mathrm{Var}[X_i] + 2 \sum_{i<j} \mathrm{Cov}(X_i, X_j).$$

The *correlation coefficient* between random variables X and Y is defined to be

$$\beta = \frac{\mathrm{Cov}(X, Y)}{\sqrt{\mathrm{Var}X}\sqrt{\mathrm{Var}Y}}.$$

It can be shown that $-1 \leq \beta \leq 1$ [34]. If $\beta > 0$, then X and Y are said to be *positively correlated*, and if $\beta < 0$, then X and Y are said to be *negatively correlated*. Loosely speaking, positively correlated random variables tend to increase or decrease together, while negatively correlated random variables tend to move in opposite directions. When $\beta = 0$, X and Y are said to be *uncorrelated*.

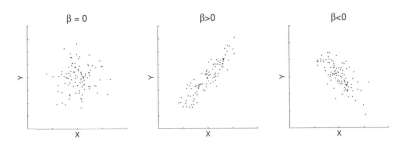

Figure 1.4: Representative samples of random vector (X, Y).

Example 1.7. A portfolio manager wishes to allocate $1 million between two assets. Denote by X_i the return of the i-th asset, and assume $E[X_i] = r_i$ and $\mathrm{Var}[X_i] = \sigma_i^2$ for $i = 1, 2$. The correlation coefficient between X_1 and X_2 is assumed to be β. It is required that the overall expected return from the allocation be no smaller than a given level r^*. The portfolio manager would like to choose such an allocation with minimal variance. Solve this *mean variance optimization* problem under the assumption that

$$r_1 = 0.2, \quad r_2 = 0.1, \quad \sigma_1^2 = 0.1, \quad \sigma_2^2 = 0.4, \quad \beta = -0.5, \quad r^* = 0.15.$$

bitch what? do these formulas come where from?

SOLUTION: Suppose that the strategy is to invest $\$w_i$ million in the i-th asset. Then,

$$w_i \geq 0, \quad w_1 + w_2 = 1.$$

The return of this strategy is $w_1 X_1 + w_2 X_2$. The expected return and variance are, respectively,

$$
\begin{aligned}
E[w_1 X_1 + w_2 X_2] &= w_1 r_1 + w_2 r_2, \\
\mathrm{Var}[w_1 X_1 + w_2 X_2] &= w_1^2 \sigma_1^2 + w_2^2 \sigma_2^2 + 2\beta \sigma_1 \sigma_2 w_1 w_2.
\end{aligned}
$$

Therefore, the optimization problem is to minimize

$$w_1^2 \sigma_1^2 + w_2^2 \sigma_2^2 + 2\beta \sigma_1 \sigma_2 w_1 w_2$$

under the constraints that

$$w_1 r_1 + w_2 r_2 \geq r^*, \quad w_1 + w_2 = 1, \quad w_i \geq 0.$$

Plugging in the given parameters and substituting $1 - w_1$ for w_2, the optimization problem reduces to minimizing

$$0.7 w_1^2 - w_1 + 0.4 \quad \text{such that} \quad 0.5 \leq w_1 \leq 1.$$

Thus, the optimal allocation is $w_1^* = 5/7$ and $w_2^* = 1 - w_1^* = 2/7$. The expected value and standard deviation of the return from this allocation are 0.1714 and 0.2070, respectively. ∎

1.4.2 Independence

Two random variables X and Y are said to be *independent* if for any $A, B \subseteq \mathbb{R}$

$$\mathbb{P}(X \in A, Y \in B) = \mathbb{P}(X \in A) \cdot \mathbb{P}(Y \in B).$$

If X and Y are discrete random variables, then X and Y are independent if and only if the joint probability mass function equals the product of the marginal probability mass functions:

↳ makes sense; one probability is not effected by the other.

$$p(x, y) = p_X(x) p_Y(y).$$

Similarly, if X and Y are jointly continuous, then X and Y are independent if and only if the joint density function equals the product of the marginal density functions:

$$f(x, y) = f_X(x) f_Y(y).$$

Lemma 1.8. *Assume that X and Y are independent. Then*

1. $E[XY] = E[X]E[Y]$,

2. $\text{Cov}(X, Y) = 0$,

3. $\text{Var}[X + Y] = \text{Var}[X] + \text{Var}[Y]$.

PROOF. Assume that X and Y are jointly continuous. Denote by $f(x, y)$ the joint density and f_X, f_Y the two marginals. It follows that

$$
\begin{aligned}
E[XY] &= \iint_{\mathbb{R}^2} xyf(x, y)\, dxdy \\
&= \iint_{\mathbb{R}^2} xyf_X(x)f_Y(y)\, dxdy \\
&= \int_{\mathbb{R}} xf_X(x)\, dx \int_{\mathbb{R}} yf_Y(y)\, dy \\
&= E[X]E[Y].
\end{aligned}
$$

(2) follows immediately from (1), and (3) is a consequence of (2) and Lemma 1.7. The proof for the discrete case is similar. ∎

A very useful result concerning normal random variables is that any linear combination of independent normal random variables is still normally distributed. *why is that useful?*

Lemma 1.9. *Assume that $\{X_1, \dots, X_n\}$ are independent and X_i is normally distributed as $N(\mu_i, \sigma_i^2)$ for each i. Then for any constants $\{a_1, \dots, a_n\}$,*

$$
\sum_{i=1}^{n} a_i X_i = N\left(\sum_{i=1}^{n} a_i \mu_i, \sum_{i=1}^{n} a_i^2 \sigma_i^2 \right).
$$

The proof can be found in many probability textbooks; e.g., [34, Chapter 6]. See also Exercise 1.14.

Example 1.8. Suppose that the payoff of an option depends on the prices of two stocks, say S_1 and S_2. If S_i exceeds k_i for each i, then the payoff is a fixed amount; otherwise, it is zero. The evaluation of this option will involve the expected value of $h(S_1, S_2)$, where

$$
h(x_1, x_2) = \begin{cases} 1 & \text{if } x_i > k_i \text{ for } i = 1, 2, \\ 0 & \text{otherwise.} \end{cases}
$$

Assume that S_1 and S_2 are independent and S_i is lognormally distributed as $\text{LogN}(\mu_i, \sigma_i^2)$. Compute this expected value.

SOLUTION: Since S_1 and S_2 are independent, we can express the expected value as

$$E[h(S_1, S_2)] = \mathbb{P}(S_1 > k_1, S_2 > k_2) = \mathbb{P}(S_1 > k_1)\mathbb{P}(S_2 > k_2).$$

Note that $\log S_i$ is $N(\mu_i, \sigma_i^2)$. It follows that for each i

$$\mathbb{P}(S_i > k_i) = \mathbb{P}(\log S_i > \log k_i) = \Phi\left(-\frac{\log k_i - \mu_i}{\sigma_i}\right).$$

Therefore,

$$E[h(S_1, S_2)] = \Phi\left(-\frac{\log k_1 - \mu_1}{\sigma_1}\right) \cdot \Phi\left(-\frac{\log k_2 - \mu_2}{\sigma_2}\right). \qquad \blacksquare$$

1.5 Conditional Distributions

Consider two random variables X and Y. We are interested in the conditional distribution of X given $Y = y$. Earlier in this chapter we have defined the conditional probability of A given B, namely,

$$\mathbb{P}(A|B) = \frac{\mathbb{P}(AB)}{\mathbb{P}(B)},$$

if $\mathbb{P}(B) > 0$. Therefore, when both X and Y are discrete, the conditional distribution of X given $Y = y$ can be defined in a straightforward manner:

$$\mathbb{P}(X = x|Y = y) = \frac{\mathbb{P}(X = x, Y = y)}{\mathbb{P}(Y = y)}.$$

Difficulty arises, however, when Y is a continuous random variable. In this case, the above definition fails automatically since $\mathbb{P}(Y = y) = 0$. Nonetheless, it is appropriate to replace the probability mass functions by the density functions to define analogously a conditional density function $f(x|y)$. The following theorem can be found in many introductory probability textbooks; see e.g., [34, Chapter 5].

Theorem 1.10. *Suppose that X and Y are continuous random variables with joint density $f(x, y)$. The conditional density of X given $Y = y$ is*

$$f(x|y) = \frac{f(x, y)}{f_Y(y)}.$$

Once the conditional density is obtained, it is very easy to express quantities related to the conditional distribution. For example, for every set $A \subseteq \mathbb{R}$,

$$\mathbb{P}(X \in A | Y = y) = \int_A f(x|y)\, dx,$$

and for any function $h : \mathbb{R} \to \mathbb{R}$,

$$E[h(X)|Y = y] = \int_{\mathbb{R}} h(x) f(x|y)\, dx.$$

Analogous to Theorem 1.4, one can write down another version of the law of total probability in terms of random variables. The proof is straightforward and thus omitted.

Theorem 1.11. (Law of Total Probability). *For any random variables X and Y and any subset* $A \subseteq \mathbb{R}$,

doesn't this need to have a number on top?

1. *Y is discrete:*

$$\mathbb{P}(X \in A) = \sum_j \mathbb{P}(X \in A | Y = y_j) \mathbb{P}(Y = y_j).$$

2. *Y is continuous:*

$$\mathbb{P}(X \in A) = \int_{\mathbb{R}} \mathbb{P}(X \in A | Y = y) f_Y(y)\, dy.$$

Remark 1.1. In the special case where X and Y are independent, the conditional distribution of X given $Y = y$ is always the distribution of X itself, regardless of y. *yeah makes sense*

what does this mean

Example 1.9. Consider a one-factor credit risk model where losses are due to the default of obligors on contractual payments. Suppose that there are m obligors and the i-th obligor defaults if and only if $X_i \geq x_i$ for some random variable X_i and given level x_i. The random variable X_i is assumed to take the form

$$X_i = \rho_i Z + \sqrt{1 - \rho_i^2}\, \varepsilon_i,$$

where $Z, \varepsilon_1, \dots, \varepsilon_m$ are independent standard normal random variables, and each ρ_i is a given constant satisfying $-1 < \rho_i < 1$. Compute the probability that none of the obligors default.

SOLUTION: Note that for each i, the distribution of X_i is $N(0,1)$, thanks to Lemma 1.9. Therefore,   *finding an area*

$$\mathbb{P}(\text{the } i\text{-th obligor does not default}) = \mathbb{P}(X_i < x_i) = \Phi(x_i).$$

Since X_i's have a common component Z, they cannot be independent unless $\rho_i = 0$ for every i. Therefore, it is in general not correct to write

$$\mathbb{P}(\text{no default}) = \prod_{i=1}^{m} \Phi(x_i). \qquad cdf$$

However, X_i's are independent given $Z = z$. This conditional independence allows us to write

$$\mathbb{P}(\text{no default} \mid Z = z) = \prod_{i=1}^{m} \mathbb{P}\left(\varepsilon_i < \frac{x_i - \rho_i z}{\sqrt{1 - \rho_i^2}}\right) = \prod_{i=1}^{m} \Phi\left(\frac{x_i - \rho_i z}{\sqrt{1 - \rho_i^2}}\right).$$

It now follows from Theorem 1.11 that

$$\mathbb{P}(\text{no default}) = \int_{\mathbb{R}} \prod_{i=1}^{m} \Phi\left(\frac{x_i - \rho_i z}{\sqrt{1 - \rho_i^2}}\right) \cdot \frac{1}{\sqrt{2\pi}} e^{-\frac{1}{2}z^2} \, dz.$$

There is no closed form formula for this probability. One often resorts to approximations or Monte Carlo simulation to produce estimates for such quantities. *So you do a Simulation when there is no Closed form* ∎ *way to estimate the area*

Example 1.10. Another popular measure of risk is the so-called *expected tail loss*. Let L be the loss of portfolio value and $a > 0$ a given loss threshold. The expected tail loss is defined as the conditional expectation

$$\bar{L}_a = E[L \mid L > a] \qquad \text{what does } a \text{ mean here?}$$

Assume that L is normally distributed as $N(\mu, \sigma^2)$. Compute $\bar{L}_a$.

SOLUTION: The key step is to derive the conditional distribution of L given $L > a$. To give a general treatment, we temporarily assume that L has an arbitrary density f. Denote by φ the conditional density. It is natural that φ should take the form

$$\varphi(x) = \begin{cases} cf(x) & \text{if } x > a, \\ 0 & \text{if } x \leq a, \end{cases}$$

where c is some constant to be determined. Since φ is a density, it must satisfy

$$\int_{-\infty}^{\infty} \varphi(x)\,dx = 1.$$

It follows that, *from where?*

$$c = \frac{1}{\int_a^\infty f(x)\,dx} = \frac{1}{\mathbb{P}(L > a)}.$$

Therefore,

$$E[L|L > a] = \int_{-\infty}^{\infty} x\varphi(x)\,dx = \frac{1}{\mathbb{P}(L > a)} \int_a^\infty xf(x)\,dx.$$

In particular, when L is normally distributed as $N(\mu, \sigma^2)$, it is not difficult to verify that $\mathbb{P}(L > a) = \Phi(-\theta)$, where $\theta = (a - \mu)/\sigma$, and

$$
\begin{aligned}
\int_a^\infty xf(x)\,dx &= \int_a^\infty x \cdot \frac{1}{\sqrt{2\pi}\sigma} e^{-\frac{1}{2\sigma^2}(x-\mu)^2}\,dx \\
&= \int_\theta^\infty (\mu + \sigma z) \cdot \frac{1}{\sqrt{2\pi}} e^{-\frac{1}{2}z^2}\,dz \\
&= \mu\Phi(-\theta) + \frac{\sigma}{\sqrt{2\pi}} e^{-\frac{1}{2}\theta^2}.
\end{aligned}
$$

Hence, the expected tail loss is

$$\bar{L}_a = E[L|L > a] = \mu + \frac{\sigma}{\sqrt{2\pi}\Phi(-\theta)} e^{-\frac{1}{2}\theta^2}. \qquad \blacksquare$$

1.6 Conditional Expectation

Consider two random variables X and Y. The *conditional expectation* of X given Y, denoted by $E[X|Y]$, is a function of Y and thus a random variable itself. $E[X|Y]$ can be determined as follows.

Step 1. Compute $E[X|Y = y]$ for every fixed value y. If one knows the conditional distribution of X given $Y = y$, then $E[X|Y = y]$ is just the expected value of this conditional distribution.

the integral of this cdf

Step 2. Regard $E[X|Y = y]$ as a function of y and write $E[X|Y = y] = f(y)$. Replace y by Y to obtain $E[X|Y]$. That is,

$$E[X|Y] = f(Y).$$

A very important result is the following *tower property* of conditional expectations.

Theorem 1.12. (Tower Property). *For any random variables X and Y,*

$$E[E[X|Y]] = E[X].$$

PROOF. We will show for the case where X and Y have a joint density function $f(x, y)$. The proof for other cases is similar and thus omitted. By the definition of conditional expectation,

$$
\begin{aligned}
E[E[X|Y]] &= \int_{\mathbb{R}} E[X|Y = y] f_Y(y) \, dy \\
&= \int_{\mathbb{R}} \left[\int_{\mathbb{R}} x f(x|y) \, dx \right] f_Y(y) \, dy \\
&= \int_{\mathbb{R}} \int_{\mathbb{R}} x f(x, y) \, dx dy \\
&= \int_{\mathbb{R}} x f_X(x) \, dx \\
&= E[X].
\end{aligned}
$$

This completes the proof.   ∎

Lemma 1.13. *Given any random variables X, Y, Z, any constants a, b, c, and any function h : $\mathbb{R} \to \mathbb{R}$,*

1. $E[c|Y] = c.$

2. $E[h(Y)|Y] = h(Y).$

3. $E[aX + bZ|Y] = aE[X|Y] + bE[Z|Y].$

4. $E[h(Y)X|Y] = h(Y)E[X|Y].$

5. $E[h(X)|Y] = E[h(X)]$, *provided that X and Y are independent.*

The proof of this lemma is left as an exercise to the reader.

Remark 1.2. Even though Y is assumed to be a random variable in our dealings with conditional distributions and conditional expectations, it can actually be much more general. For example, Y can be any random vectors. All the results that we have stated hold for the general case.

Example 1.11. Stock prices are sometimes modeled by distributions other than lognormal in order to fit the empirical data more accurately. For instance, Merton [22] introduced a jump diffusion model for stock prices. A special case of Merton's model assumes that the underlying stock price S satisfies

$$S = e^Y, \quad Y = X_1 + \sum_{i=1}^{X_2} Z_i,$$

where X_1 is $N(\mu, \sigma^2)$, X_2 is Poisson with parameter λ, Z_i is $N(0, v^2)$, and $X_1, X_2, \{Z_i\}$ are all independent. The evaluation of call options involves expected values such as

$$E[(S - K)^+],$$

where K is some positive constant. Compute this expected value.

SOLUTION: For every $n \geq 0$, we can compute the conditional expected value

$$v_n = E[(S - K)^+ | X_2 = n].$$

Indeed, conditional on $X_2 = n$, Y is normally distributed as $N(\mu, \sigma^2 + nv^2)$, and thus S is lognormally distributed with parameters μ and $\sigma^2 + nv^2$. It follows from Example 1.6 that

$$v_n = e^{\mu + \frac{1}{2}(\sigma^2 + nv^2)} \Phi\left(\sqrt{\sigma^2 + nv^2} - \theta_n\right) - K\Phi(-\theta_n), \quad \theta_n = \frac{\log K - \mu}{\sqrt{\sigma^2 + nv^2}}.$$

By Theorem 1.12 or the tower property of conditional expectations,

$$E[(S - K)^+] = E[E[(S - K)^+ | X_2]] = \sum_{n=0}^{\infty} v_n \mathbb{P}(X_2 = n) = e^{-\lambda} \sum_{n=0}^{\infty} v_n \frac{\lambda^n}{n!}.$$

This can be evaluated numerically. ∎

1.7 Classical Limit Theorems

Two of the most important limit theorems in probability theory are the strong law of large numbers and central limit theorem. They have numerous applications in both theory and practice. In particular, they provide the theoretical foundation for Monte Carlo simulation schemes.

A rigorous proof of these two theorems is beyond the scope of this book and can be found in a number of advanced probability textbooks such as [2, 5]. We will simply state the theorems without proof. In what follows, the acronym "iid" stands for "independent identically distributed."

Theorem 1.14. (Strong Law of Large Numbers). *If $X_1, X_2, \cdots$ are iid random variables with mean μ, then*

$$\mathbb{P}\left\{\frac{X_1 + X_2 + \cdots + X_n}{n} \to \mu\right\} = 1.$$

Theorem 1.15. (Central Limit Theorem). *If $X_1, X_2, \cdots$ are iid random variables with mean μ and variance σ^2, then the distribution of*

$$\frac{X_1 + X_2 + \cdots + X_n - n\mu}{\sigma\sqrt{n}}$$

converges to the standard normal distribution. That is, for any $a \in \mathbb{R}$,

$$\mathbb{P}\left\{\frac{X_1 + X_2 + \cdots + X_n - n\mu}{\sigma\sqrt{n}} \le a\right\} \to \int_{-\infty}^{a} \frac{1}{\sqrt{2\pi}} e^{-\frac{1}{2}x^2}\, dx$$

as $n \to \infty$.

Exercises

Pen-and-Paper Problems

1.1 Let A and B be two events such that $A \subseteq B$. Show that $\mathbb{P}(A) \leq \mathbb{P}(B)$.

1.2 Let A and B be two arbitrary events. Show that $\mathbb{P}(A \cup B) \leq \mathbb{P}(A) + \mathbb{P}(B)$.

1.3 Let X be a random variable and let a and b be two arbitrary constants. Show that
$$E[aX + b] = aE[X] + b, \quad \text{Var}[aX + b] = a^2 \text{Var}[X].$$

1.4 Let X be a random variable that is uniformly distributed on $(0, 1)$. Show that for any constants $a < b$,
$$Y = a + (b - a)X$$
is uniformly distributed on (a, b).

1.5 Show that the density function of the lognormal distribution $\text{LogN}(\mu, \sigma^2)$ is given by
$$f(x) = \frac{1}{x\sigma\sqrt{2\pi}} e^{-\frac{1}{2\sigma^2}(\log x - \mu)^2}, \quad x > 0.$$

1.6 Assume that X has distribution $N(\mu, \sigma^2)$. Let a and b be two arbitrary constants and $Y = a + bX$.

    (a) Find the cumulative distribution function of Y in terms of Φ.

    (b) Compute the density of Y and show that Y is normally distributed with mean $a + b\mu$ and variance $b^2\sigma^2$.

1.7 Assume that X is a standard normal random variable. For an arbitrary $\theta \in \mathbb{R}$, show that $E[\exp\{\theta X\}] = \exp\{\theta^2/2\}$.

1.8 Assume that X is normally distributed with mean μ and variance σ^2. For an arbitrary $\theta \in \mathbb{R}$, determine $E[\exp\{\theta X\}]$.

1.9 Suppose that X is a lognormal random variable with distribution $\text{LogN}(\mu, \sigma^2)$.

    (a) Given a constant $a > 0$, what is the distribution of aX?

    (b) Given a constant $\alpha \in \mathbb{R}$, what is the distribution of X^α?

1.10 Suppose that S is lognormally distributed as $\text{LogN}(\mu, \sigma^2)$. Compute

    (a) $E[(K - S)^+]$ and $E[\max\{S, K\}]$, where K is a positive constant;

    (b) $E[(S^\alpha - K)]^+$, where α and K are given positive constants.

1.11 Suppose that the joint default probability distribution of two bonds A and B is as follows.

	Bond A	
Bond B	Default	No default
Default	0.05	0.10
No default	0.05	0.80

Joint default probability distribution

(a) What is the default probability of bond A?

(b) What is the default probability of bond B?

(c) Given that bond A defaults, what is the probability that bond B defaults?

(d) Are the defaults of bond A and bond B independent, positively correlated, or negatively correlated?

1.12 Let S_1 and S_2 be the prices of two assets. Assume that $X = \log S_1$ and $Y = \log S_2$ have a joint density function

$$f(x,y) = \frac{\sqrt{3}}{4\pi} e^{-\frac{1}{2}(x^2 - xy + y^2)}, \quad \text{for } x, y \in \mathbb{R}.$$

(a) What is the distribution of X?

(b) What is the distribution of Y?

(c) What is the distribution of X given $Y = y$?

(d) Determine $E[X|Y]$.

(e) Determine $E[S_1|S_2]$ and use tower property to compute $E[S_1 S_2]$.

(f) Compute $\text{Cov}(S_1, S_2)$.

The random vector (X, Y) is an example of jointly normal random vectors; see Appendix A.

1.13 Let X be a standard normal random variable. Assume that given $X = x$, Y is normally distributed as $N(x, 1)$. Find $\text{Cov}(X, Y)$. *Hint:* Use tower property to compute $E[Y]$ and $E[XY]$.

1.14 Assume that X and Y are two independent random variables. Show that

(a) if X is binomial with parameters (n, p) and Y is binomial with parameters (m, p), then $X + Y$ is binomial with parameters $(m + n, p)$;

(b) if X is Poisson with parameter λ and Y is Poisson with parameter μ, then $X + Y$ is Poisson with parameter $\lambda + \mu$;

(c) if X has cumulative distribution function F and Y has density g, then for any $z \in \mathbb{R}$

$$\mathbb{P}(X + Y \leq z) = \int_{\mathbb{R}} F(z - y) g(y) \, dy;$$

(d) if X has density f and Y has density g, then the density of $X + Y$ is given by the *convolution* $f * g$, where

$$(f * g)(z) = \int_{\mathbb{R}} f(z - y)g(y)\, dy = \int_{\mathbb{R}} g(z - x)f(x)\, dx;$$

(e) if X is normally distributed as $N(0, \sigma_1^2)$ and Y is normally distributed as $N(0, \sigma_2^2)$, then $X + Y$ is normally distributed as $N(0, \sigma_1^2 + \sigma_2^2)$;

(f) if X is exponential with rate λ and Y is exponential with rate μ, then $\min\{X, Y\}$ is exponential with rate $\lambda + \mu$.

1.15 Assume that $X_1, \ldots, X_n$ are independent random variables and that for each i, X_i is lognormal with distribution $\text{LogN}(\mu_i, \sigma_i^2)$. Find the distribution of the product random variable $X_1 \cdot X_2 \cdots \cdot X_n$.

1.16 Denote by X_i the change of a portfolio's value at the i-th day. Assume that $\{X_i\}$ are independent and identically distributed normal random variables with distribution $N(\mu, \sigma^2)$. Determine the value-at-risk at the confidence level $1 - \alpha$ for the total loss over an m-day period.

1.17 Suppose that L, the loss of a portfolio's value, is lognormally distributed with parameters μ and σ^2. Given a constant $a > 0$, compute the expected tail loss

$$E[L|L > a].$$

1.18 The concept of *utility maximization* plays an important role in the study of economics and finance. The basic setup is as follows. Let X be the return from an investment. The distribution of X varies according to the investment strategy employed. The goal is to maximize the expected utility $E[U(X)]$ by judiciously picking a strategy, for some given *utility function U* that is often assumed to be concave and increasing.

Consider the following utility maximization problem with the utility function $U(x) = \log x$. An investor has a \$1 million capital and can choose to invest any portion of it. Suppose that y is the amount invested. Then with probability p the amount invested will double, and with probability $1 - p$ the amount invested will be lost. Let X be the total wealth in the end. Assuming $p > 0.5$, determine the value of y that maximizes the expected utility

$$E[\log X].$$

1.19 Assume that $X_1, \ldots, X_n$ are iid random variables with mean μ and variance σ^2. Define

$$\bar{X} = \frac{1}{n} \sum_{i=1}^{n} X_i, \quad S^2 = \frac{1}{n-1} \sum_{i=1}^{n} (X_i - \bar{X})^2.$$

Show that $E[\bar{X}] = \mu$ and $E[S^2] = \sigma^2$. $\bar{X}$ and S^2 are standard sample estimates for μ and σ^2, respectively.

1.20 Assume that $(X_1, Y_1), \ldots, (X_n, Y_n)$ are iid random vectors. Define

$$R = \frac{1}{n-1} \sum_{i=1}^{n} (X_i - \bar{X})(Y_i - \bar{Y}), \quad \bar{X} = \frac{1}{n} \sum_{i=1}^{n} X_i, \quad \bar{Y} = \frac{1}{n} \sum_{i=1}^{n} Y_i.$$

Show that $E[R] = \mathrm{Cov}(X_i, Y_i)$ for every i.

1.21 Let $X_1, \ldots, X_n$ be iid random variables with common density $f_\theta(x)$, where θ is an unknown parameter. The joint density of $(X_1, \ldots, X_n)$ is

$$\prod_{i=1}^{n} f_\theta(x_i).$$

The *maximum likelihood estimate* of θ is defined to be

$$\hat{\theta} = \mathrm{argmax}_\theta \prod_{i=1}^{n} f_\theta(X_i).$$

Assume that $X_1, \ldots, X_n$ are iid normal random variables with distribution $N(\mu, \sigma^2)$. Show that the maximum likelihood estimate of $\theta = (\mu, \sigma^2)$ is given by

$$\hat{\mu} = \bar{X} = \frac{1}{n} \sum_{i=1}^{n} X_i, \quad \hat{\sigma}^2 = \frac{1}{n} \sum_{i=1}^{n} (X_i - \bar{X})^2.$$

1.22 Silver Wheaton (SLW) is a mining company that generates its revenue primarily from the sales of silver. iShares Silver Trust (SLV) is a grantor trust that provides a vehicle for investors to own interests in silver. The table below contains the stock prices of SLW and SLV from Oct 24, 2011 to Nov 4, 2011.

Day i	1	2	3	4	5
SLW$_i$	31.24	32.21	33.40	34.46	35.97
SLV$_i$	30.86	32.42	32.50	32.92	34.27
Day i	6	7	8	9	10
SLW$_i$	34.60	34.07	34.91	36.05	36.09
SLV$_i$	33.44	32.33	32.23	33.61	33.17

Stock prices of SLW and SLV

Define $X_i = \log(\mathrm{SLW}_{i+1}/\mathrm{SLW}_i)$ and $Y_i = \log(\mathrm{SLV}_{i+1}/\mathrm{SLV}_i)$ for $i = 1, \ldots, 9$. Assume that (X_i, Y_i) are iid random vectors. Use the results in Exercise 1.19 and Exercise 1.20 to estimate the correlation coefficient between X and Y.

1.23 A function $f : \mathbb{R} \to \mathbb{R}$ is said to be *convex* if for any $\lambda \in [0, 1]$ and $x_1, x_2 \in \mathbb{R}$,

$$f(\lambda x_1 + (1 - \lambda)x_2) \leq \lambda f(x_1) + (1 - \lambda)f(x_2).$$

Let X be an arbitrary random variable with $\mu = E[X]$. Assume that f is differentiable at $x = \mu$. Show that for any x

$$f(x) \geq f(\mu) + f'(\mu)(x - \mu),$$

and thus

$$E[f(X)] \geq f(E[X]).$$

This inequality is called the *Jensen's inequality*.

MATLAB® Problems

The three commands, "normcdf", "norminv", and "normrnd", are MATLAB functions relevant to normal distributions. To get the detailed description of, say, "norminv", you can use the MATLAB command "help norminv."

"normcdf(x)": return $\Phi(x)$.

"norminv(α)": return the value of x such that $\Phi(x) = \alpha$.

"normrnd": generate samples from normal distributions.

1.A Compute the value-at-risk in Exercise 1.16 for $\mu = 0.1$, $\sigma = 1$, $m = 10$, and $\alpha = 0.01$.

1.B Compute the expected tail loss in Exercise 1.17 for $\mu = 0$, $\sigma = 1$, and $a = 2$.

1.C Generate 10 samples from each of the following distributions.

(a) The standard normal distribution.

(b) The lognormal distribution with parameters $\mu = 1$ and $\sigma^2 = 4$.

(c) The binomial distribution with parameters $(20, 0.5)$. Use "binornd".

(d) The uniform distribution on $(5, 7)$. Use "rand".

(e) The Poisson distribution with parameter 1. Use "poissrnd".

(f) The exponential distribution with rate 2. Use "exprnd".

1.D Use the command "plot" to plot the density functions of the exponential distributions with rate $\lambda = 1$, $\lambda = 0.5$, and $\lambda = 2$, respectively. Use a different line type for each density curve and use the command "legend" to place a legend on the picture, similar to Figure 1.1.

1.E Suppose that X is a standard normal random variable and given $X = x$, Y is normally distributed with mean x and variance one. Generate 1000 samples from the joint distribution of (X, Y) and use the command "scatter" to plot your samples. Are X and Y positively correlated or negatively correlated? See also Exercise 1.13.

1.F Repeat Exercise 1.E except that, given $X = x$, Y is normally distributed with mean $-x$ and variance one. By looking at the plot of the samples, do you expect X and Y to be positively or negatively related? Verify your answer by computing the theoretical value of the correlation coefficient between X and Y.

Chapter 2

Brownian Motion

Brownian motion was discovered in 1827 by the English botanist Robert Brown when he was studying the movement of microscopic pollen grains suspended in a drop of water. The rigorous mathematical foundation of Brownian motion was established by Norbert Wiener around 1923. For this reason, Brownian motion is also called the Wiener process. In mathematical finance, Brownian motion has been used extensively in the modeling of security prices. The celebrated Black–Scholes option pricing formula was derived upon the assumption that the underlying stock price is a geometric Brownian motion.

   The purpose of this chapter is to introduce Brownian motion and its basic properties. We suggest that the reader go over Appendix A before reading this chapter since the multivariate normal distributions are indispensable for the study of Brownian motion.

2.1   Brownian Motion

A continuous time stochastic process $W = \{W_t : t \geq 0\}$ is a collection of random variables indexed by "time" t. For each fixed $\omega \in \Omega$ the mapping $t \mapsto W_t(\omega)$ is called a *sample path*. We say W is a **standard Brownian motion** if the following conditions hold:

1. Every sample path of the process W is continuous.

2. $W_0 = 0$.

3. The process has independent increments, that is, for any sequence

$0 = t_0 < t_1 < \cdots < t_n$, the increments

$$W_{t_1} - W_{t_0}, \ W_{t_2} - W_{t_1}, \ \cdots, \ W_{t_n} - W_{t_{n-1}}$$

are independent random variables.

4. For any $s \geq 0$ and $t > 0$, the increment $W_{s+t} - W_s$ is normally distributed with mean 0 and variance t.

It is immediate from the definition that $W_t = W_t - W_0$ is normally distributed with mean 0 and variance t.

Figure 2.1 shows some representative sample paths of a standard Brownian motion. They all exhibit a certain kind of ruggedness. Actually, it can be shown that with probability one, the Brownian motion sample paths are nowhere differentiable and nowhere monotonic [18].

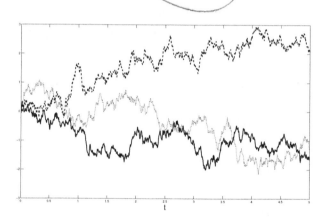

Figure 2.1: Sample paths of Brownian motion.

The next lemma follows directly from the definition of Brownian motion. We leave the proof to the reader.

Lemma 2.1. *Suppose that $W = \{W_t : t \geq 0\}$ is a standard Brownian motion. Then the following statements hold.*

1. *(Symmetry) The process $-W = \{-W_t : t \geq 0\}$ is a standard Brownian motion.*

2. *Fix an arbitrary $s > 0$ and define $B_t = W_{t+s} - W_s$ for $t \geq 0$. Then $B = \{B_t : t \geq 0\}$ is a standard Brownian motion.*

2.2 Running Maximum of Brownian Motion

The *running maximum* of a standard Brownian motion W by time t is defined to be

$$M_t = \max_{0 \le s \le t} W_s.$$

It is possible to derive analytically the distribution of M_t, as well as the joint distribution of (W_t, M_t), through the so-called *reflection principle* of Brownian motion.

To illustrate, consider a fixed level $b > 0$ and define the first passage time of the Brownian motion to the level b:

$$T_b = \inf\{t \ge 0 : W_t = b\}.$$

Note that T_b is random and

$$\mathbb{P}(M_t \ge b) = \mathbb{P}(T_b \le t).$$

The reflection principle asserts that

$$\mathbb{P}(W_t \le b \mid T_b \le t) = \mathbb{P}(W_t \ge b \mid T_b \le t) = \frac{1}{2}. \tag{2.1}$$

The intuition is as follows. Lemma 2.1(2) basically states that a Brownian

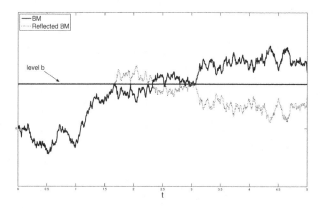

Figure 2.2: Reflected Brownian motion.

motion starts afresh at any deterministic time s. With a leap of faith, assume that it also starts afresh at the random time T_b. Therefore, given $T_b \le t$, by the symmetry of Brownian motion, for every path that reaches a point above b at time t there is a "reflected" path that reaches a point below b at time t; see the solid path and its reflection, which is represented by the dotted path, in Figure 2.2. Since $\{W_t \geq b\} \subseteq \{T_b \leq t\}$, it follows that

$$\mathbb{P}(W_t \geq b \mid T_b \leq t) = \frac{\mathbb{P}(W_t \geq b, T_b \leq t)}{\mathbb{P}(T_b \leq t)} = \frac{\mathbb{P}(W_t \geq b)}{\mathbb{P}(T_b \leq t)}.$$

Therefore, by (2.1)

$$P(M_t \geq b) = \mathbb{P}(T_b \leq t) = 2\mathbb{P}(W_t \geq b) = 2\Phi\left(-\frac{b}{\sqrt{t}}\right).$$

Taking derivatives with respect to b on both sides, we arrive at the density of M_t:

what is M_t?

$$f(x) = \frac{2}{\sqrt{2\pi t}} e^{-\frac{1}{2t}x^2} \tag{2.2}$$

for $x \geq 0$.

As for the joint density function of (W_t, M_t), fix arbitrarily $a \leq b$ and $b > 0$. Analogous to (2.1), we have

$$\mathbb{P}(W_t \leq a \mid T_b \leq t) = \mathbb{P}(W_t \geq 2b - a \mid T_b \leq t).$$

Therefore,

$$
\begin{aligned}
\mathbb{P}(W_t \leq a, M_t \geq b) &= \mathbb{P}(W_t \leq a, T_b \leq t) \\
&= \mathbb{P}(W_t \geq 2b - a, T_b \leq t) \\
&= \mathbb{P}(W_t \geq 2b - a) \\
&= \Phi\left(-\frac{2b - a}{\sqrt{t}}\right).
\end{aligned}
$$

Taking derivatives over a and b on both sides, it follows that the joint density function of (W_t, M_t) is

$$f(x, y) = \frac{2(2y - x)}{\sqrt{2\pi t^3}} e^{-\frac{1}{2t}(2y - x)^2} \tag{2.3}$$

for $x \leq y$ and $y \geq 0$.

Remark 2.1. The claim that a Brownian motion starts afresh at the first passage time T_b is a consequence of the *strong Markov property*. The proof is advanced and rather technical; see [18].

2.3   Derivatives and Black–Scholes Prices

Financial derivatives such as options derive their values from the underlying assets. For example, a call option on a prescribed stock with strike price K and maturity T grants the holder of the option the right to buy the stock at the price K at time T. Denote by S_t the stock price at time t. If at maturity the stock price S_T is above K, the holder can exercise the option, namely, buy the stock at the price K and sell it immediately at the market price S_T, to realize a profit of $S_T - K$. If the stock price S_T is at or below K, then the option expires worthless. In other words, the payoff of this call option at time T is

$$(S_T - K)^+.$$

In general, a financial derivative pays a random amount X at a given maturity date T. The form of the payoff X can be very simple and only depends *diff types* on the price of the underlying asset at time T such as call options, or it *of options* can be very complicated and depend on the entire history of the asset price up until time T. The most fundamental question in the theory of financial derivatives is about evaluation: what is the fair price of a derivative with payoff X and maturity T?

To answer the question, we need a model for the price of the underlying asset. The classical Black–Scholes model assumes that the asset price is a *geometric Brownian motion*, that is,

$$S_t = S_0 \exp\left\{\left(\mu - \frac{1}{2}\sigma^2\right)t + \sigma W_t\right\}, \quad \text{mean? variance?} \tag{2.4}$$

where $W = \{W_t : t \geq 0\}$ is a standard Brownian motion, and S_0 is the initial or current asset price. The pair of parameters μ and $\sigma > 0$ are said to be the *drift* and *volatility*, respectively. In contrast to a Brownian motion, a geometric Brownian motion is always positive. Furthermore, for every $t > 0$, S_t is lognormally distributed with distribution

the price of the stock can

$$\text{LogN}(\log S_0 + (\mu - \sigma^2/2)t, \sigma^2 t). \quad \text{fall anywhere across this distribution.}$$

Many pricing problems admit explicit solutions when the underlying asset price is modeled by a geometric Brownian motion.

In this section, we evaluate a number of financial derivatives, assuming that the price of the underlying asset is the geometric Brownian motion (2.4) with drift

$$\mu = r, \tag{2.5}$$

where r denotes the *risk-free interest rate*. Throughout the book, the interest rate is always assumed to be continuously compounded. In practice, the risk-free interest rate is usually taken to be the yield on a zero-coupon bond with similar maturity.

The price or the value of a financial derivative with payoff X and maturity T is given by

$$v = E[e^{-rT}X]. \tag{2.6}$$

Therefore, pricing a financial derivative amounts to computing an expected value. It is not difficult to understand the discounting factor e^{-rT} in the pricing formula (2.6), since one dollar at time T is worth e^{-rT} dollars at time 0. The real question is why we *equate* the drift of the underlying asset price with the risk-free interest rate r. This indeed follows from the arbitrage free principle, which is the topic of the next chapter. For the time being, we assume that the condition (2.5) and the pricing formula (2.6) are valid and use them to evaluate some derivatives. We repeat that, unless otherwise specified, the price of the underlying asset is assumed to be a geometric Brownian motion with drift r and volatility σ.

Example 2.1. (The Black–Scholes Formula). A call option with strike price K and maturity T has payoff $(S_T - K)^+$ at time T. Price this option.

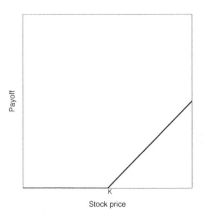

Figure 2.3: Payoff of call option with strike price K.

SOLUTION: Thanks to equation (2.6), the price of the call option is given by

$$v = E[e^{-rT}(S_T - K)^+].$$

Since S_T is lognormally distributed as $\text{LogN}(\log S_0 + (r - \sigma^2/2)T, \sigma^2 T)$, it follows from Example 1.6 that

$$v = S_0 \Phi(\sigma\sqrt{T} - \theta) - K e^{-rT} \Phi(-\theta),$$

where

$$\theta = \frac{1}{\sigma\sqrt{T}} \log \frac{K}{S_0} + \left(\frac{\sigma}{2} - \frac{r}{\sigma}\right) \sqrt{T}.$$

The call option price was first derived in the seminal paper by Fischer Black and Myron Scholes [3]. For future reference, we will denote the call option price v by

$$\text{BLS\_Call}(S_0, K, T, r, \sigma).$$

Sometimes we simply use "BLS_Call" when there is no confusion about the parameters. ∎

Example 2.2. A put option with strike price K and maturity T gives the holder of the option the right to sell the stock at the price K at maturity T. Its payoff is $(K - S_T)^+$. Determine its price.

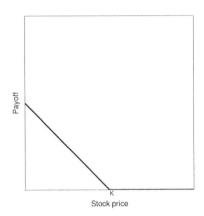

Figure 2.4: Payoff of put option with strike price K.

SOLUTION: Analogous to Example 2.1, we should denote the price of the put option by

$$\text{BLS\_Put}(S_0, K, T, r, \sigma),$$

or simply "BLS_Put" when there is no confusion about the parameters. Observe that

$$(S_T - K)^+ - (K - S_T)^+ = S_T - K.$$

Therefore, thanks to (2.6) and Exercise 2.9, we arrive at the *put-call parity*, namely,

$$\text{BLS\_Call} - \text{BLS\_Put} = E[e^{-rT}(S_T - K)] = S_0 - e^{-rT}K.$$

It follows that

$$\text{BLS\_Put} = e^{-rT}K\Phi(\theta) - S_0\Phi(\theta - \sigma\sqrt{T}),$$

where θ is as defined in Example 2.1. ∎

Example 2.3. A binary call option with maturity T pays one dollar when the stock price at time T is at or above a certain level K and pays nothing otherwise. The payoff can be written in the form of an indicator function:

$$X = 1_{\{S_T \geq K\}}.$$

Compute the price of this option.

SOLUTION: It is trivial that $E[X] = \mathbb{P}(S_T \geq K)$. Therefore, the price of the option is

$$v = e^{-rT}\mathbb{P}(S_T \geq K).$$

Since $\log S_T$ is normally distributed with mean $\log S_0 + (r - \sigma^2/2)T$ and variance $\sigma^2 T$, it follows that

$$v = e^{-rT}\mathbb{P}(\log S_T \geq \log K) = e^{-rT}\Phi\left(-\frac{\log(K/S_0) - (r - \sigma^2/2)T}{\sigma\sqrt{T}}\right). \quad \blacksquare$$

Example 2.4. The buyer of a *future* contract is obliged to buy the underlying asset at a certain price p at a specified future time T. Unlike options, upon entering the contract, the buyer does not need to pay any premium. However, at time T the contract must be honored. Therefore, the payoff to the buyer at time T is

$$X = S_T - p.$$

The future price p is chosen so that the value of the contract at present is zero. What should be the value of p?

SOLUTION: The value of the contract to the buyer at present is

$$E[e^{-rT}X] = E[e^{-rT}S_T] - e^{-rT}p = S_0 - e^{-rT}p.$$

Hence, for the contract to have value zero one must have

$$p = e^{rT} S_0.$$

Clearly, with this choice of p the value of the contract is also zero to the seller of the contract. ∎

Example 2.5. An Asian option is a path-dependent option whose payoff depends on the average of the underlying stock price over the option life. Consider a discretely monitored average price call option with payoff

$$X = (\bar{S} - K)^+$$

at maturity T, where $\bar{S}$ is the geometric average of the stock price defined by

$$\bar{S} = \left(\prod_{i=1}^{m} S_{t_i} \right)^{1/m}$$

for a given set of monitoring dates $0 \le t_1 < \cdots < t_m \le T$. Calculate the price of this option.

SOLUTION: It is not difficult to see that $\bar{S}$ is lognormally distributed. Indeed,

$$\log \bar{S} = \frac{1}{m} \sum_{i=1}^{m} \log S_{t_i} = \log S_0 + \frac{1}{m} \sum_{i=1}^{m} \left[\left(r - \frac{\sigma^2}{2} \right) t_i + \sigma W_{t_i} \right]$$

is a linear transform of the jointly normal random vector $(W_{t_1}, \ldots, W_{t_m})$ and hence normal itself (see Appendix A and Exercise 2.6). The mean and the variance of $\log \bar{S}$ are

$$
\begin{aligned}
\bar{\mu} &= \log S_0 + \left(r - \frac{\sigma^2}{2} \right) \bar{t}, \quad \text{where } \bar{t} = \frac{1}{m} \sum_{i=1}^{m} t_i, \\
\bar{\sigma}^2 &= \mathrm{Var} \left(\frac{1}{m} \sum_{i=1}^{m} \sigma W_{t_i} \right) \\
&= \frac{\sigma^2}{m^2} \left[\sum_{i=1}^{m} \mathrm{Var}(W_{t_i}) + 2 \sum_{i<j} \mathrm{Cov}(W_{t_i}, W_{t_j}) \right] \\
&= \frac{\sigma^2}{m^2} \left[\sum_{i=1}^{m} t_i + 2 \sum_{i<j} t_i \right] \\
&= \frac{\sigma^2}{m^2} \sum_{i=1}^{m} (2m - 2i + 1) t_i.
\end{aligned}
$$

Therefore, the price of this Asian option is, thanks to Example 1.6,

$$v = E[e^{-rT}X] = e^{-rT}\left[e^{\bar{\mu}+\frac{1}{2}\bar{\sigma}^2}\Phi(\bar{\sigma}-\theta) - K\Phi(-\theta)\right]$$

with $\theta = (\log K - \bar{\mu})/\bar{\sigma}$.   ∎

The next example is about pricing a lookback call option. The computation relies on Lemma 2.2, which is a preliminary version of the Girsanov's Theorem [18]. Note that we say a function is *path-dependent* if it depends on the sample paths of the relevant process. For example, if we define

$$h(W_{[0,T]}) = \max_{0\leq t\leq T} W_t - \min_{0\leq t\leq T} W_t - W_T,$$

then h is a path-dependent function and its value depends on the entire sample path $W_{[0,T]} = \{W_t : 0 \leq t \leq T\}$.

Lemma 2.2. *Given an arbitrary constant θ, let $B = \{B_t : t \geq 0\}$ be a Brownian motion with drift θ. That is,*

$$B_t = W_t + \theta t, \quad t \geq 0,$$

where $W = \{W_t : t \geq 0\}$ is a standard Brownian motion. Then for any $T > 0$ and path-dependent function h,

$$E\left[h(B_{[0,T]})\right] = E\left[e^{\theta W_T - \frac{1}{2}\theta^2 T}h(W_{[0,T]})\right].$$

PROOF. Since W_T is normally distributed as $N(0,T)$, it follows from the tower property that the right-hand-side equals

$$\text{R.H.S.} = \int_{\mathbb{R}} E\left[e^{\theta W_T - \frac{1}{2}\theta^2 T}h(W_{[0,T]})\Big| W_T = x\right] \cdot \frac{1}{\sqrt{2\pi T}}e^{-\frac{1}{2T}x^2}\,dx$$

$$= \int_{\mathbb{R}} E\left[h(W_{[0,T]})\Big| W_T = x\right] \cdot f(x)\,dx, \tag{2.7}$$

where

$$f(x) = e^{\theta x - \frac{1}{2}\theta^2 T} \cdot \frac{1}{\sqrt{2\pi T}}e^{-\frac{1}{2T}x^2} = \frac{1}{\sqrt{2\pi T}}e^{-\frac{1}{2T}(x-\theta T)^2}.$$

On the other hand, Exercise 2.12 shows that the conditional distribution of $\{B_t : 0 \leq t \leq T\}$ given $B_T = x$ is the same as that of $\{W_t : 0 \leq t \leq T\}$ given $W_T = x$. Therefore,

$$E\left[h(W_{[0,T]})\Big| W_T = x\right] = E\left[h(B_{[0,T]})\Big| B_T = x\right].$$

Since $f(x)$ is indeed the density of B_T, the tower property implies that the integral in (2.7) equals

$$E\left[E\left[h(B_{[0,T]})\big|\,B_T\right]\right] = E\left[h(B_{[0,T]})\right].$$

This completes the proof. ∎

Example 2.6. A lookback call option with fixed strike price K and maturity T is a path-dependent option, whose payoff is

$$X = \left(\max_{0\le t\le T} S_t - K\right)^+.$$

Assuming $K > S_0$, what is the value of this option at time zero?

SOLUTION: The value of the option is $v = E[e^{-rT}X]$. To compute this expected value, define

$$B_t = W_t + \theta t, \quad \theta = \left(\frac{r}{\sigma} - \frac{\sigma}{2}\right)t.$$

Then,

$$X = \left(\max_{0\le t\le T} S_t - K\right)^+ = \left(S_0 \exp\left\{\sigma \cdot \max_{0\le t\le T} B_t\right\} - K\right)^+.$$

By Lemma 2.2,

$$E[X] = E\left[e^{\theta W_T - \frac{1}{2}\theta^2 T}\left(S_0 e^{\sigma \cdot M_T} - K\right)^+\right], \quad M_T = \max_{0\le t\le T} W_t.$$

It follows that

$$E[X] = \int_0^\infty \int_{-\infty}^\infty e^{\theta x - \frac{1}{2}\theta^2 T}\left(S_0 e^{\sigma y} - K\right)^+ \cdot f(x,y)\,dxdy,$$

where $f(x,y)$ is the joint density of (W_T, M_T) given by (2.3) with $t = T$. The evaluation of this double integral is rather straightforward but tedious. We will only state the result and leave the details to the reader (Exercise 2.13 may prove useful for this endeavor):

$$\begin{aligned}
v &= \text{BLS\_Call}(S_0, K, T, r, \sigma) \\
&\quad + \frac{\sigma^2}{2r} S_0\left[\Phi(\theta_+) - e^{-rT}\left(\frac{K}{S_0}\right)^{2r/\sigma^2}\Phi(\theta_-)\right],
\end{aligned}$$

where

$$\theta_{\pm} = \frac{1}{\sigma\sqrt{T}} \log \frac{S_0}{K} + \left(\frac{\sigma}{2} \pm \frac{r}{\sigma}\right)\sqrt{T}.$$

Lemma 2.2 can also be used to evaluate other similar path-dependent options such as lookback options with floating strike price and barrier options. See Exercise 2.14. ∎

Example 2.7. Consider two stocks whose prices are modeled by geometric Brownian motions:

$$S_t = S_0 \exp\left\{\left(r - \frac{\sigma_1^2}{2}\right)t + \sigma_1 W_t\right\},$$

$$V_t = V_0 \exp\left\{\left(r - \frac{\sigma_2^2}{2}\right)t + \sigma_2 B_t\right\}.$$

For simplicity, we assume that W and B are two independent standard Brownian motions. The payoff of an exchange option at maturity T is

$$(S_T - V_T)^+.$$

Compute the price of this option.

SOLUTION: The price of this option is

$$v = E[e^{-rT}(S_T - V_T)^+] = E\left[e^{-rT}V_T\left(\frac{S_T}{V_T} - 1\right)^+\right].$$

Straightforward computation yields that

$$e^{-rT}V_T = V_0 \exp\left\{-\frac{\sigma_2^2}{2}T + \sigma_2 B_T\right\},$$

$$\frac{S_T}{V_T} = \frac{S_0}{V_0}\exp\left\{\frac{\sigma_2^2 - \sigma_1^2}{2}T - \sigma_2 B_T + \sigma_1 W_T\right\}.$$

Since W_T and B_T are independent $N(0, T)$ random variables, their joint density function equals

$$f(x, y) = \frac{1}{\sqrt{2\pi T}}\exp\left\{-\frac{x^2}{2T}\right\}\frac{1}{\sqrt{2\pi T}}\exp\left\{-\frac{y^2}{2T}\right\}.$$

Therefore, one can express v in terms of an integral with respect to the density f and obtain

$$v = V_0 \iint_{\mathbb{R}^2} \left[\frac{S_0}{V_0} \exp\left\{ \frac{\sigma_2^2 - \sigma_1^2}{2} T - \sigma_2 y + \sigma_1 x \right\} - 1 \right]^+ g(x, y) \, dx dy,$$

where

$$
\begin{aligned}
g(x, y) &= f(x, y) \exp\left\{ -\frac{\sigma_2^2}{2} T + \sigma_2 y \right\} \\
&= \frac{1}{\sqrt{2\pi T}} \exp\left\{ -\frac{x^2}{2T} \right\} \frac{1}{\sqrt{2\pi T}} \exp\left\{ -\frac{(y - \sigma_2 T)^2}{2T} \right\}.
\end{aligned}
$$

It is interesting to observe that $g(x, y)$ itself is the *joint density function* of two independent normal random variables, say X and Y, where X is $N(0, T)$ and Y is $N(\sigma_2 T, T)$. Therefore, the price v can be written as

$$
\begin{aligned}
v &= V_0 E\left[\left(\frac{S_0}{V_0} \exp\left\{ \frac{\sigma_2^2 - \sigma_1^2}{2} T - \sigma_2 Y + \sigma_1 X \right\} - 1 \right)^+ \right] \\
&= V_0 E\left[(U - 1)^+ \right],
\end{aligned}
$$

where U stands for a lognormal random variable with distribution

$$\text{LogN} \left(\log \frac{S_0}{V_0} - \frac{\sigma_2^2 + \sigma_1^2}{2} T, \ \sigma_2^2 T + \sigma_1^2 T \right).$$

Letting $\sigma = \sqrt{\sigma_1^2 + \sigma_2^2}$, it follows from Example 1.6 that

$$v = S_0 \Phi \left(\frac{1}{\sigma\sqrt{T}} \log \frac{S_0}{V_0} + \frac{\sigma\sqrt{T}}{2} \right) - V_0 \Phi \left(\frac{1}{\sigma\sqrt{T}} \log \frac{S_0}{V_0} - \frac{\sigma\sqrt{T}}{2} \right).$$

The trick in this calculation is to value of the option in terms of the stock price V_T. This technique is called *change of numéraire*. It is very useful in the evaluation of option prices. ∎

2.4 Multidimensional Brownian Motions

For the purpose of future reference, we give the definition of a multidimensional Brownian motion.

Consider a continuous time process $B = \{B_t : t \geq 0\}$ where B_t is a d-dimensional random vector for each t. Let $\Sigma = [\Sigma_{ij}]$ be a $d \times d$ symmetric positive definite matrix. The process B is said to be a *d-dimensional Brownian motion with covariance matrix Σ* if the following conditions are satisfied:

1. Every sample path of the process B is continuous.

2. $B_0 = 0$.

3. The process has independent increments, that is, for any sequence $0 = t_0 < t_1 < \cdots < t_n$, the increments

$$B_{t_1} - B_{t_0}, \ B_{t_2} - B_{t_1}, \ \cdots, \ B_{t_n} - B_{t_{n-1}}$$

are independent random vectors.

4. For any $s \geq 0$ and $t > 0$, the increment $B_{s+t} - B_s$ is a jointly normal random vector with mean 0 and covariance matrix $t\Sigma$.

It follows immediately from the definition that B_t is a d-dimensional jointly normal random vector with distribution $N(0, t\Sigma)$.

When $\Sigma = I_d$, we say B is a *d-dimensional standard Brownian motion*. In this case, every component of B is a one-dimensional standard Brownian motion itself, and all these components are independent. Note that in general, the components of a d-dimensional Brownian motion may not be independent. Two components, say the i-th component and the j-th component, are independent if and only if $\Sigma_{ij} = 0$.

Exercises

Pen-and-Paper Problems

2.1 Assume that $X = (X_1, \ldots, X_d)'$ is a d-dimensional normal random vector with distribution $N(0, \Sigma)$. Let A be an invertible matrix such that

$$AA' = \Sigma.$$

Find the distribution of $Y = A^{-1}X$.

2.2 Suppose that a portfolio consists of two assets and the change of portfolio's value, denoted by X, can be written as

$$X = \beta X_1 + (1 - \beta) X_2,$$

where X_i denotes the change of value of the i-th asset and $0 \leq \beta \leq 1$. Assume that (X_1, X_2) is a jointly normal random vector with distribution

$$N\left(\begin{bmatrix} \mu \\ \mu \end{bmatrix}, \begin{bmatrix} 1 & \rho \\ \rho & 1 \end{bmatrix} \right).$$

(a) Find the distribution of X.

(b) What is the value-at-risk at the confidence level $1 - \alpha$ for the total loss of the portfolio?

(c) For what value of β will the value-at-risk be minimized?

2.3 Suppose that a portfolio consists of two assets and the total value of the portfolio is

$$Y = \beta S_1 + (1 - \beta) S_2,$$

where β is the allocation parameter and $0 \leq \beta \leq 1$. Assume that

$$S_1 = e^{X_1}, \quad S_2 = e^{X_2},$$

where (X_1, X_2) is a jointly normal random vector with distribution

$$N\left(\begin{bmatrix} 0 \\ 0 \end{bmatrix}, \begin{bmatrix} 1 & 0 \\ 0 & \sigma^2 \end{bmatrix} \right).$$

(a) Compute $\mathrm{Var}[S_1]$ and $\mathrm{Var}[S_2]$.

(b) Determine the optimal allocation β that minimizes $\mathrm{Var}[Y]$.

2.4 Brownian motion can be regarded as the limit of simple random walks. Consider a sequence of iid random variable $\{X_i\}$ such that

$$\mathbb{P}(X_i = 1) = \frac{1}{2} = \mathbb{P}(X_i = -1).$$

Fix an arbitrary n. Let $t_m = m/n$ for $m = 0, 1, \ldots$ and define a discrete-time stochastic process $W^{(n)} = \{W^{(n)}_{t_m} : m = 0, 1, \ldots\}$ where

$$W^{(n)}_{t_m} = \frac{1}{\sqrt{n}} \sum_{i=1}^{m} X_i.$$

Show that $W^{(n)}$ has independent and stationary increments, that is, the increments

$$W^{(n)}_{t_1} - W^{(n)}_{t_0}, \cdots, W^{(n)}_{t_m} - W^{(n)}_{t_{m-1}}, \cdots$$

are iid. Assume that $t_m \to t + h$ and $t_k \to t$ as $n \to \infty$. Use central limit theorem to determine the limit distribution of

$$W^{(n)}_{t_m} - W^{(n)}_{t_k}$$

as $n \to \infty$. Explain intuitively that $W^{(n)}$ converges to a standard Brownian motion.

2.5 Suppose that W is a standard Brownian motion. Given an arbitrary constant $a > 0$, show that $B = \{B_t : t \geq 0\}$, where

$$B_t = \frac{1}{\sqrt{a}} W_{at},$$

is also a standard Brownian motion.

2.6 Suppose that W is a standard Brownian motion. Show that the following statements hold.

(a) The conditional distribution of W_t given $W_s = x$ is $N(x, t - s)$ for any $0 \leq s < t$.

(b) For $0 < s < t$, the conditional distribution of W_s given $W_t = x$ is $N(xs/t, s(t - s)/t)$.

(c) The covariance of W_s and W_t is $s \wedge t = \min\{s, t\}$ for any $s, t \geq 0$.

(d) For any $0 < t_1 < t_2 < \cdots < t_n$, the random vector $(W_{t_1}, \ldots, W_{t_n})'$ is jointly normal with distribution $N(0, \Sigma)$, where $\Sigma = [\sigma_{ij}]$ is an $n \times n$ matrix with $\sigma_{ij} = t_i \wedge t_j$.

(e) For any $t \geq 0$ and $\theta \in \mathbb{R}$, $E[\exp\{\theta W_t\}] = \exp\{\theta^2 t/2\}$.

2.7 Let W be a standard Brownian motion. Given any $0 = t_0 < t_1 < \cdots < t_m$ and constants $a_1, a_2, \ldots, a_m$, find the distributions of $\sum_{i=1}^{m} a_i(W_{t_i} - W_{t_{i-1}})$ and $\sum_{i=1}^{m} a_i W_{t_i}$.

2.8 Let W be a standard Brownian motion. For any $t > 0$, determine the density of the *running minimum* by time t

$$m_t = \min_{0 \le s \le t} W_s$$

and the joint density of (W_t, m_t).

2.9 Let $S = \{S_t : t \ge 0\}$ be a geometric Brownian motion with drift μ and volatility σ. Given any $T \ge 0$, show that

$$E[S_T] = e^{\mu T} S_0.$$

2.10 Show that the Black–Scholes call option price BLS_Call(S_0, K, T, r, σ) is monotonically increasing with respect to the volatility σ.

2.11 Assume that the price of an underlying asset is a geometric Brownian motion

$$S_t = S_0 \exp\left\{\left(r - \frac{1}{2}\sigma^2\right)t + \sigma W_t\right\},$$

where r is the risk-free interest rate. Determine the prices of the following derivatives with maturity T and payoff X, in terms of the Black–Scholes call option price formula "BLS_Call".

    (a) Break forwards $X = \max\{S_T, S_0 e^{rT}\}$.
    (b) Collar option $X = \min\{\max\{S_T, K_1\}, K_2\}$ with $0 < K_1 < K_2$.
    (c) Forward-start option $X = (S_T - S_{T_0})^+$ with $T_0 < T$.
    (d) Straddle

$$X = \begin{cases} K - S_T & \text{if } S_T \le K, \\ S_T - K & \text{if } S_T \ge K. \end{cases}$$

    (e) Power call option $(S_T^\beta - K)^+$ for some positive constant β.

2.12 Let W be a standard Brownian motion. The process $B = \{B_t : t \ge 0\}$, where $B_t = W_t + \theta t$ for some constant θ, is said to be a *Brownian motion with drift* θ. Given $0 < t_1 < \cdots < t_n < T$, show that the conditional distribution of $(B_{t_1}, \ldots, B_{t_n})$ given $B_T = x$ does not depend on θ. In particular, letting $\theta = 0$, we conclude that the conditional distribution of $\{B_t : 0 \le t \le T\}$ given $B_T = x$ is the same as the conditional distribution of $\{W_t : 0 \le t \le T\}$ given $W_T = x$.

2.13 By convention, let Φ denote the cumulative distribution function of the standard normal distribution. Given any constants $a \ne 0$ and θ, use integration by parts to show that

$$\begin{aligned}
\int_\theta^\infty e^{ax} \Phi(-x)\, dx &= -\frac{1}{a} e^{a\theta} \Phi(-\theta) + \int_\theta^\infty \frac{1}{a} e^{ax} \frac{1}{\sqrt{2\pi}} e^{-\frac{1}{2}x^2}\, dx \\
&= -\frac{1}{a} e^{a\theta} \Phi(-\theta) + \frac{1}{a} e^{\frac{1}{2}a^2} \Phi(-\theta + a).
\end{aligned}$$

Similarly, when $a = 0$, show that

$$\int_\theta^\infty \Phi(-x)\,dx = -\theta\Phi(-\theta) + \int_\theta^\infty x\frac{1}{\sqrt{2\pi}}e^{-\frac{1}{2}x^2}dx$$
$$= -\theta\Phi(-\theta) + \frac{1}{\sqrt{2\pi}}e^{-\frac{1}{2}\theta^2}.$$

2.14 Use Lemma 2.2 to determine the prices of the following path-dependent options with maturity T and payoff X, assuming that the underlying stock price is a geometric Brownian motion with drift r and volatility σ.

(a) Lookback put option with floating strike price: $X = \max_{0 \le t \le T} S_t - S_T$.

(b) Lookback put option with fixed strike price K: $X = (K - \min_{0 \le t \le T} S_t)^+$.

(c) Down-and-out call option with strike price K and barrier b (assume $S_0 > b$):
$$X = (S_T - K)^+ \cdot 1_{\{\min_{0 \le t \le T} S_t \ge b\}}.$$

(d) Up-and-in call option with strike price K and barrier b (assume $S_0 < b$):
$$X = (S_T - K)^+ \cdot 1_{\{\max_{0 \le t \le T} S_t \ge b\}}.$$

2.15 Suppose that W is a d-dimensional Brownian motion with covariance matrix Σ. Let A be an $m \times d$ matrix and define

$$B_t = AW_t.$$

Show that B is an m-dimensional Brownian motion with covariance matrix $A\Sigma A'$.

2.16 Suppose that (W, B) is a two-dimensional Brownian motion with covariance matrix

$$\begin{bmatrix} 1 & \rho \\ \rho & 1 \end{bmatrix}$$

for some $\rho \in (-1, 1)$. Show that (Q, B) is a two-dimensional standard Brownian motion, where

$$Q_t = \frac{1}{\sqrt{1-\rho^2}}(W_t - \rho B_t).$$

In particular, Q and B are independent standard Brownian motions.

2.17 Brownian motion is the fundamental continuous time continuous process. On the other hand, the fundamental continuous time jump process is Poisson process. One way to construct a *Poisson process* $N = \{N_t : t \ge 0\}$ *with intensity* λ is as follows. Let $\{X_n\}$ be a sequence of iid exponential random variable with rate λ. Let $S_0 = 0$ and for $n \ge 1$

$$S_n = \sum_{i=1}^n X_i.$$

Define for any $t \geq 0$

$$N_t = k, \quad \text{if } S_k \leq t < S_{k+1}.$$

Show that

(a) N has independent increments, that is, for any n and $0 = t_0 < t_1 < \cdots < t_n$, the increments

$$N_{t_1} - N_{t_0}, \cdots, N_{t_n} - N_{t_{n-1}}$$

are independent;

(b) $N_{t+s} - N_s$ is a Poisson random variable with parameter λt for any $s \geq 0$ and $t > 0$. In particular, N_t is a Poisson random variable with parameter λt.

A very useful observation for analyzing Poisson processes is the *memoryless property* of exponential distributions. That is, for any $t, s \geq 0$,

$$\mathbb{P}(X > t + s | X > t) = \mathbb{P}(X > s)$$

when X is exponentially distributed.

MATLAB® Problems

2.A Write a function using the "function" command to calculate the price of a call option from the Black–Scholes formula, namely BLS_Call. The function should have input parameters

$$
\begin{aligned}
r &= \text{risk-free interest rate,} \\
\sigma &= \text{volatility,} \\
T &= \text{maturity,} \\
K &= \text{strike price,} \\
S_0 &= \text{initial stock price.}
\end{aligned}
$$

The output of the function is the Black–Scholes price of the corresponding call option. Save this function as a ".m" file.

(a) Compute the price of a call option with $S_0 = 50$, $r = 0.05$, $\sigma = 0.2$, $T = 0.5$, and strike price $K = 45, 50, 55$, respectively.

(b) Consider a call option with $S_0 = 50$, $r = 0.05$, $K = 50$, and $T = 0.5$. Plot the price of this call option with respect to the volatility σ for $0 < \sigma \leq 0.5$. Suppose that at present the market price of the call option is $3. Find the value of the volatility from your plot such that

$$\text{BLS\_Call}(50, 50, 0.5, 0.05, \sigma) = \text{market price } \$3.$$

This volatility is said to be the *implied volatility*.

(c) Consider a call option with $S_0 = 50$, $r = 0.05$, $K = 50$, and $\sigma = 0.2$. Plot the price of this call option with respect to the maturity T for $0 \leq T \leq 1$. Is this an increasing function?

2.B Write a function using the "function" command to calculate the price of a put option from the Black–Scholes formula, namely BLS_Put. The function should have input parameters

$$
\begin{aligned}
r &= \text{risk-free interest rate,} \\
\sigma &= \text{volatility,} \\
T &= \text{maturity,} \\
K &= \text{strike price,} \\
S_0 &= \text{initial stock price.}
\end{aligned}
$$

The output of the function is the Black–Scholes price of the corresponding put option. Save this function as a ".m" file.

(a) Compute the price of a put option with $S_0 = 50$, $r = 0.05$, $\sigma = 0.2$, $T = 0.5$, and strike price $K = 45, 50, 55$, respectively.

(b) Consider a put option with $K = 50$, $r = 0.05$, $\sigma = 0.2$, and $T = 0.5$. Plot the price of this put option with respect to the initial stock price S_0 for $40 \leq S_0 \leq 60$. Is it an increasing function or decreasing function?

(c) Consider a put option with $S_0 = 50$, $r = 0.05$, $K = 50$, and $\sigma = 0.2$. Plot the price of this put option with respect to the maturity T for $0 \leq T \leq 1$. Is it an increasing function or decreasing function?

(d) Consider a put option with $S_0 = 50$, $r = 0.05$, $K = 50$, and $T = 0.5$. Plot the price of this put option with respect to the volatility σ for $0 < \sigma \leq 0.5$. Is it an increasing function or decreasing function? Suppose that at present the market price of the put option is \$2. Find the implied volatility from your plot. That is, determine the value of the volatility such that

$$
\text{BLS\_Put}(50, 50, 0.5, 0.05, \sigma) = \text{market price \$2.}
$$

Chapter 3

Arbitrage Free Pricing

A question from Chapter 2 remains open: why is it appropriate to equate the drift of the stock price with the risk-free interest rate r? It should be emphasized that this question is meaningful only in the context of pricing financial derivatives. It is clearly not true if we consider the real world stock price movements—in general one would expect the drift or the growth rate of a stock to be higher than the risk-free interest rate due to the risk associated with the stock.

The purpose of this chapter is to explain the key idea—arbitrage free principle—behind the pricing of financial derivatives. We will work with the binomial tree asset pricing models, not only because they are widely used in practice, but also because they provide probably the simplest setting to illustrate the mechanism of arbitrage free pricing. The answer to the open question from Chapter 2 becomes transparent once one realizes that a geometric Brownian motion can be approximated by binomial trees.

3.1 Arbitrage Free Principle

Consider a call option with strike price K and maturity T. The real world price of the underlying asset is assumed to be a geometric Brownian motion with drift μ and volatility σ:

$$S_t = S_0 \exp\left\{\left(\mu - \frac{\sigma^2}{2}\right)t + \sigma W_t\right\},$$

where $W = \{W_t : t \geq 0\}$ is a standard Brownian motion. It is tempting to set the price of the call option as the expected value of its discounted payoff

$$E[e^{-rT}(S_T - K)^+]. \tag{3.1}$$

If this were correct, then the price of a call option would be higher if the underlying asset has a higher growth rate or drift, everything else being equal.

Unfortunately, this intuitive approach is *not* correct. As we will demonstrate later, if the option price is determined in this fashion, then one can construct portfolios that generate *arbitrage* opportunities. By arbitrage, we mean a portfolio whose value process $X = \{X_t\}$ satisfies $X_0 = 0$ and

$$\mathbb{P}(X_T \geq 0) = 1, \quad \mathbb{P}(X_T > 0) > 0.$$

The *arbitrage free principle* stipulates that there are no arbitrage opportunities or free lunch in a financial market. This principle is not far from truth. In real life, market sometimes does exhibit arbitrage. But it cannot sustain itself and will only last for a very short amount of time—as soon as it is discovered and exploited, it is removed from the market.

It turns out that under appropriate market conditions, for a financial derivative the only price that is consistent with the arbitrage free principle is the expected value of the discounted payoff, where the expected value is taken as if the drift of the underlying asset price were equal to the risk-free interest rate. That is, the right pricing formula for (say) a call option with strike price K and maturity T should still be (9.5). But instead of a geometric Brownian with drift μ, the price of the underlying asset is treated as a geometric Brownian motion with drift r:

$$S_t = S_0 \exp\left\{\left(r - \frac{\sigma^2}{2}\right)t + \sigma W_t\right\}.$$

Therefore, the option price should *not* depend on the true drift of the underlying asset.

Since the price is derived upon the arbitrage free principle, it is called the *arbitrage free price*. But what is the motivation for arbitrage free pricing? In principle, asset prices are the results of the equilibrium between demand and supply. This equilibrium pricing approach is often used in economics to study the prices of assets *endogenously*. However, to successfully carry it out, it is necessary to characterize the preference and the attitude toward risk for each of the participating agents in the market. Clearly this is not very practical. Therefore, in financial engineering a different and more practical approach is adopted. Assuming that the prices of a collection of assets such as stocks are given *exogenously*, one tries to determine the prices of other assets such as options based on the assumption that the market is free of arbitrage. In this sense, arbitrage free pricing is a relative pricing

* prices are given exogenously?

mechanism. A thorough investigation on derivative pricing can be found in [7, 16].

Unless otherwise specified, the financial derivatives under consideration are assumed to be *European*, that is, they can only be exercised at the maturity date. On the contrary, if a derivative can be exercised at any time before or at the maturity date, it is said to be *American*. The value of an American financial derivative is obviously at least as much as that of its European counterpart. See Appendix B for the pricing of American derivatives.

Remark 3.1. The pricing mechanism is sometimes called the *risk-neutral pricing*, since the expected value is taken in an artificial world where all the risky assets have the same growth rate as the risk free saving account, or equivalently, all the investors are risk neutral.

Remark 3.2. Throughout the book, we assume that the financial market is frictionless in the sense that there is neither any restriction nor any transaction cost on buying/selling any number of financial instruments.

3.2 Asset Pricing with Binomial Trees

In a binomial tree asset pricing model, the price of the underlying asset evolves as follows. If the asset price at the current time step is S, then at the next time step the price will move up to uS with probability p and move down to dS with probability $q = 1 - p$. Here u and d are given positive constants such that $d < u$. For all binomial tree models, we use S_n to denote the asset price at the n-th time step.

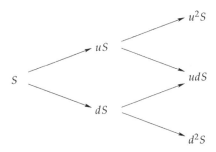

Figure 3.1: A binomial tree model.

Even though a binomial tree model seems to be too simplistic compared with the real world stock price dynamics, it proves to be a reasonable approximation in many occasions and has superior computational tractability. Figure 3.1 depicts a two-period binomial tree model.

3.2.1 A Preliminary Example

In some sense, arbitrage free pricing is a *deterministic* pricing theory made out of probabilistic models. To be more concrete, consider a one-period binomial tree model where the current stock price is $S_0 = 10$ and $u = 1/d = 2$. The risk-free interest rate r is taken to be 0 for convenience. Consider also a call option with strike price $K = 14$ and maturity $T = 1$. The option payoff at time $T = 1$ is

$$X = (S_1 - 14)^+.$$

What should be the price or the value of this option at time 0?

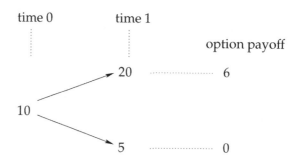

Figure 3.2: A one-period binomial pricing model.

Arbitrage Free Pricing: Consider a portfolio that consists of one share of the call option and x shares of the underlying stock. Note that the value of the portfolio at $T = 1$ is either $20x + 6$ if the stock price climbs to 20 or $5x$ if the stock price drops to 5. Therefore, if we pick x so that

$$20x + 6 = 5x \quad \text{or} \quad x = -0.4,$$

then the value of the portfolio is *fixed* at $20x + 6 = 5x = -2$ at maturity, *no matter how the stock price moves.*

Suppose that the call option's price is v at $t = 0$. Then the value of the portfolio at $t = 0$ is $10x + v = -4 + v$. Since the interest rate r equals 0, we expect that

$$-4 + v = -2 \quad \text{or} \quad v = 2.$$

That is, the option is worth $2 at time 0. Indeed, if $v \neq 2$, one can construct portfolios that lead to arbitrage:

1. $v > 2$. In this case, the option is over valued. Starting with zero initial wealth, sell one share of the call option and buy 0.4 shares of the stock, which yields a cash position of $v - 4$. At maturity, the cash position is still $v - 4$ since $r = 0$. Now selling the stocks and honoring the call option will always yield $2 no matter what the stock price is at maturity. Therefore, the total value of the portfolio becomes $v - 2 > 0$ at time $T = 1$. This is an arbitrage.

2. $v < 2$. In this case, the option is under valued. Starting with zero initial wealth, buy one share of the call option and sell 0.4 shares of the stock, which yields a cash position of $4 - v$. At maturity, the cash position is still $4 - v$ since $r = 0$. Now fulfilling the short position in the stock and exercising the call option will always yield $-$2 no matter what the stock price is at maturity. Therefore, the total value of the portfolio becomes $2 - v > 0$ at time $T = 1$. This is again an arbitrage.

3.2.2 General Pricing Formula

Now consider a general one-period binomial tree model with $S_0 = S$. We would like to price an option with maturity $T = 1$ and payoff

$$X = \begin{cases} C_u & \text{if } S_1 = uS, \\ C_d & \text{if } S_1 = dS; \end{cases}$$

see Figure 3.3. Suppose that one dollar at time $t = 0$ is worth R dollars at time $t = 1$. It is required by the arbitrage free principle (see Exercise 3.1) that

$$d < R < u.$$

Construct a portfolio with one share of the option and x shares of the underlying stock so that the portfolio will have a fixed value at maturity, regardless of the stock price. That is,

$$uS \cdot x + C_u = dS \cdot x + C_d \quad \text{or} \quad x = -\frac{C_u - C_d}{uS - dS}.$$

Suppose that the price of the option is v at time $t = 0$. Then the portfolio's value at time $t = 0$ is $v + xS$. Since one dollar at time $t = 0$ is worth R

dollars at time $t = 1$, it follows that

$$(v + xS)R = uS \cdot x + C_u,$$

which implies

$$v = \frac{1}{R}\left(\frac{R-d}{u-d} \cdot C_u + \frac{u-R}{u-d} \cdot C_d\right).$$

The reader is suggested to mimic the example in Section 3.2.1 and verify the existence of arbitrage opportunities when the option is priced differently. Note that the parameters p and q play *no* role in the pricing of the option.

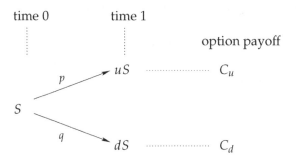

Figure 3.3: A general one-period binomial pricing model.

Risk-Neutral Probability: Observe that the price of the option can be written as the expected value of the discounted option payoff under the risk-neutral probability

$$(p^*, q^*) = \left(\frac{R-d}{u-d}, \frac{u-R}{u-d}\right). \tag{3.2}$$

That is, the option price is

$$v = E\left[\frac{1}{R}X\right] = \frac{1}{R}(p^*C_u + q^*C_d),$$

where the expected value is taken as if the stock price would move up to uS with probability p^* and move down to dS with probability $q^* = 1 - p^*$; see Figure 3.4.

Call — right to buy
put — right to sell 57

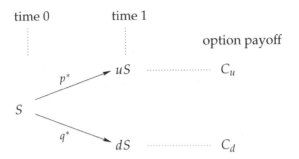

Figure 3.4: Binomial tree under the risk-neutral probability.

Summary.

1. The real world probabilities (p, q) play no role in the pricing of options.

2. The price of an option is the expected value of the discounted option payoff at maturity, under the risk-neutral probability.

3. The risk-neutral probability is determined by the structure of the binomial tree model and the interest rate, and is independent of the option payoff.

4. Under the risk-neutral probability, the growth rate of the underlying asset price equals the risk-free interest rate:

$$E[S_1] = p^* \cdot uS + q^* \cdot dS = \left(\frac{R-d}{u-d} \cdot uS + \frac{u-R}{u-d} \cdot dS \right) = RS.$$

5. **Replication:** Starting with an initial wealth that equals the price of the option, one can construct a portfolio that consists of cash and the underlying asset, and whose value at maturity completely replicates the option payoff. Indeed, starting with v dollars, one can buy

$$\delta = \frac{C_u - C_d}{uS - dS}$$

shares of the underlying asset at time $t = 0$. The cash position at time 0 will become $v - \delta S$. At time $t = 1$, the value of the portfolio is

$$R(v - \delta S) + \delta S_1 = \begin{cases} C_u & \text{if } S_1 = uS, \\ C_d & \text{if } S_1 = dS. \end{cases}$$

3.2.3   Multiperiod Binomial Tree Models

The generalization to multiperiod binomial tree models is straightforward. Consider a binomial tree with n periods and an option with maturity $T = n$ and payoff X. The parameters $u, d,$ and R are defined as before.

The conclusion is that the price of this option at time $t = 0$ is the expected value of the discounted option payoff

$$v = E\left[\frac{1}{R^n}X\right],\tag{3.3}$$

where the expected value is taken under the risk-neutral probability measure. That is, the stock price has probability p^* to move up by a factor u and probability $q^* = 1 - p^*$ to move down by a factor d at each time step. The probabilities (p^*, q^*) are given by (3.2).

In order to verify the pricing formula (3.3), and at the same time introduce a useful recursive algorithm for computing the expected value, we specialize to a three-period binomial tree model as depicted in Figure 3.5.

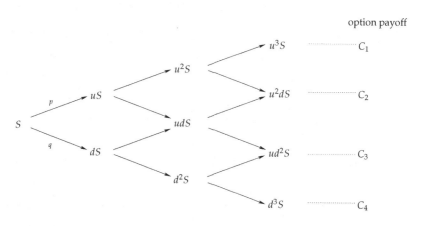

Figure 3.5: A three-period binomial tree model.

Now define V_i to be the value of the option at the i-th time step. See Figure 3.6. In general, V_i is a function of the stock price S_i and thus a random variable itself. For example, $V_2 = a$ when $S_2 = u^2 S$ and $V_1 = e$ when $S_1 = dS$. The values of V_i, or $\{a, b, \cdots, f\}$, can be obtained recursively backwards in time.

1. **At time $t = 2$:**

$$a = \frac{1}{R}(p^*C_1 + q^*C_2), \; b = \frac{1}{R}(p^*C_2 + q^*C_3), \; c = \frac{1}{R}(p^*C_3 + q^*C_4).$$

2. **At time $t = 1$:**

$$d = \frac{1}{R}(p^*a + q^*b), \ e = \frac{1}{R}(p^*b + q^*c).$$

3. **At time $t = 0$:**

$$f = \frac{1}{R}(p^*d + q^*e).$$

To summarize, the option's value V satisfies the backwards recursive equation

$$V_3 = X, \ V_i = E\left[\frac{1}{R}V_{i+1}\bigg| S_i\right], \ \ i = 2,1,0.$$

It follows from the tower property (Theorem 1.12) that for each $i = 0,1,2$

$$E[V_i] = E\left[\frac{1}{R}V_{i+1}\right].$$

Therefore,

$$f = V_0 = E\left[\frac{1}{R^3}V_3\right] = E\left[\frac{1}{R^3}X\right].$$

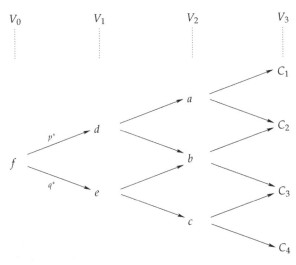

Figure 3.6: The value process of the option.

Observe that with an initial wealth of V_0, one can construct a portfolio consisting of cash and stock, and whose value at time $t = 3$ completely replicates the payoff of the option. Indeed, at time $t = 0$, it is possible to construct a portfolio whose value at $t = 1$ is exactly V_1, as described in the summary of Section 3.2.2. In a completely analogous fashion, one can adjust the portfolio at $t = 1$ (the adjustment should depend on whether $S_1 = uS$ or $S_1 = dS$) so that the value of the portfolio will replicate V_2 at $t = 2$, and so on. Now it follows immediately that V_0 is the only price of the option that is consistent with the arbitrage free principle. Indeed, if the price of the option is v and $v \neq V_0$, then one can construct arbitrage opportunities:

1. $v > V_0$: sell one share of the option and buy such a replicating portfolio.

2. $v < V_0$: buy one share of the option and sell such a replicating portfolio.

This justifies the pricing formula (3.3) for $n = 3$. The treatment for a general multiperiod binomial tree model is completely analogous.

Example 3.1. Consider a binomial tree model where each step represents 2 months in real time and $u = 1.1, d = 0.9, S_0 = 20$.

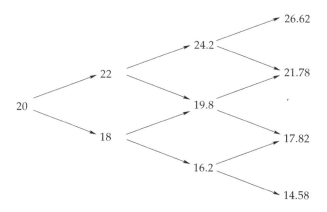

Suppose that the risk-free interest rate is 12% per annum. Compute the price of a call option with maturity $T = 6$ months and strike price $K = 21$.

SOLUTION: The risk-free interest rate for each 2-month period is $0.12/6 = 0.02$. Therefore, the discounting factor is

$$\frac{1}{R} = e^{-0.02} = 0.98,$$

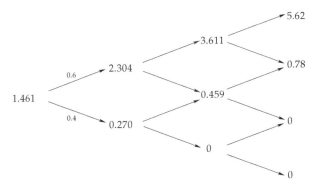

Figure 3.7: The value of the option.

and the risk-neutral probability measure is given by

$$p^* = \frac{R-d}{u-d} = 0.6, \quad q^* = \frac{u-R}{u-d} = 0.4.$$

The price of this call option at time $t = 0$ is 1.461; see Figure 3.7. The value of the option at each node is obtained backwards in time. ∎

3.3 The Black–Scholes Model

Consider the Black–Scholes model where the price of the underlying stock is assumed to be a geometric Brownian motion

$$S_t = S_0 \exp\left\{\left(\mu - \frac{\sigma^2}{2}\right) t + \sigma W_t\right\}. \tag{3.4}$$

The goal is to explain why one should replace the drift μ by the risk-free interest rate r when it comes to pricing derivatives.

The idea is to use binomial trees to approximate the geometric Brownian motion. There are many ways to achieve this approximation. We will use the following version.

1. *Binomial approximation*. Suppose that the maturity date is T. Divide the time interval $[0, T]$ into n pieces of equal length $\Delta t = T/n$. Eventually we will send n to infinity. Consider an approximating binomial tree with n periods, where each period corresponds to Δt in real time. The stock price will move up by a factor u with probability p and move down by a factor d with probability $q = 1 - p$ at each time period. The parameters p, u, and d will be determined so as to match

the distribution of the increments of the geometric Brownian motion $\{S_t\}$ over a time interval of length Δt. Even though it is impossible to make a complete match, the idea is to at least match the expected value and the variance. This leads to

$$
\left[
\begin{aligned}
Se^{\mu \Delta t} &= p \cdot uS + q \cdot dS, \\
S^2 e^{(2\mu + \sigma^2)\Delta t} &= p \cdot (uS)^2 + q \cdot (dS)^2.
\end{aligned}
\right.
$$

In order to solve for the three unknowns we impose a nonessential condition

$$
u = \frac{1}{d}.
$$

From these three equations we obtain (see Remark 3.3)

$$
u = e^{\sigma \sqrt{\Delta t}}, \quad d = e^{-\sigma \sqrt{\Delta t}}, \quad p = \frac{e^{\mu \Delta t} - d}{u - d}. \tag{3.5}
$$

It can be argued that this binomial tree approximates the geometric Brownian motion S as $n \to \infty$ [18, Invariance Principle of Donsker]. See also Exercise 3.5.

2. *Risk-neutral probability.* Since each time period in the binomial tree corresponds to Δt in real time,

$$
R = e^{r\Delta t}.
$$

Therefore, the risk-neutral probabilities (p^*, q^*) are given by

$$
p^* = \frac{R - d}{u - d} = \frac{e^{r\Delta t} - d}{u - d}, \quad q^* = 1 - p^*.
$$

The price of an option with payoff X and maturity T is

$$
v = E\left[\frac{1}{R^n} X\right] = E[e^{-rT} X],
$$

where the expected value is taken under the risk-neutral probability measure.

3. *The dynamics of $\{S_t\}$ under the risk-neutral probability measure.* Comparing the formula (3.5) with the formula of p^*, the only difference is that μ is replaced by r. Therefore, under the risk-neutral probability measure, the binomial tree approximates a geometric Brownian motion with drift r and volatility σ. In other words, as the limit of the binomial trees, the stock price $\{S_t\}$ is a geometric Brownian motion with drift r and volatility σ under the risk-neutral probability measure.

Summary: The price of an option with payoff X and maturity T is the expected value of the discounted option payoff, i.e.,

$$v = E[e^{-rT}X].$$

The expected value is taken with respect to the risk-neutral probability measure, under which the stock price is a geometric Brownian motion with drift r:

$$S_t = S_0 \exp\left\{\left(r - \frac{\sigma^2}{2}\right)t + \sigma W_t\right\}.$$

The above discussion also suggests a binomial tree approximation for $\{S_t\}$ under the risk-neutral probability measure: divide the time interval $[0, T]$ into n subintervals of equal length $\Delta t = T/n$ and set

$$u = e^{\sigma\sqrt{\Delta t}}, \quad d = e^{-\sigma\sqrt{\Delta t}}, \quad p^* = \frac{e^{r\Delta t} - d}{u - d}.$$

Remark 3.3. The solution is only approximate and not exact. One can check that the expected value is perfectly matched, but the variance is only matched up to order Δt. It turns out that a perfect match is not necessary, and the approximation is still valid with the choice of parameters in (3.5).

Exercises

Pen-and-Paper Problems

3.1 Show that the arbitrage free principle implies $d < R < u$ for a binomial tree asset pricing model.

3.2 Suppose that the risk-free interest rate is $r = 10\%$ per annum. Price the following financial derivatives using a two-period binomial tree with $u = 1.1$ and $d = 0.9$. The initial price of the underlying asset is assumed to be $S_0 = 50$.

(a) A call option with strike price $K = 48$ and maturity $T = 2$ months

(b) A put option with strike price $K = 50$ and maturity $T = 3$ months

(c) A vertical spread with payoff $X = (S_T - 48)^+ - (S_T - 52)^+$ and maturity $T = 6$ months

(d) A straddle with payoff $X = (50 - S_T)^+ + (S_T - 50)^+$ and maturity $T = 4$ months

3.3 Consider a put option with maturity $T = 6$ months and strike price $K = 19$. The initial price of the underlying stock is $S_0 = 20$. Assume that the risk-free interest rate is $r = 12\%$ annually. We would like to approximate the price of the put option with different binomial trees. For all the binomial trees we assume that $u = 1.1$ and $d = 0.9$.

(a) Price the put option using a one-period binomial tree.

(b) Starting with an initial wealth equal to the price you have computed in part (a), compose a portfolio that consists of cash and stock to completely replicate the payoff of the put option.

(c) Price the put option using a two-period binomial tree, each period representing 3 months in real time.

(d) Starting with an initial wealth equal to the price you have computed in part (c), compose a portfolio that consists of cash and stock to completely replicate the payoff of the put option. Determine the following.

   i. The initial cash and stock positions at time $t = 0$

   ii. The cash and stock positions at time $t = 1$ when the stock price goes up to uS_0

   iii. The cash and stock positions at time $t = 1$ when the stock price moves down to dS_0

3.4 Consider a call option with maturity $T = 6$ months and strike price $K = 19$. The initial price of the underlying stock is $S_0 = 20$. We would like to approximate the price of this call option using a two-period binomial tree with $u = 1.1$ and $d = 0.9$. In the first period, the risk-free interest rate is $r_1 = 6\%$ per annum, but in the second period the risk-free interest rate becomes $r_2 = 8\%$ per annum. Price this option and construct a replicating portfolio consisting of cash and stock.

3.5 The intuitive justification that the binomial tree defined by (3.5) approximates the geometric Brownian motion (3.4) is very similar to Exercise 2.4. The binomial tree has n periods, each period representing $\Delta t = T/n$ in real time. Denote by B_m the value of the tree at the m-th period for $0 \leq m \leq n$. Clearly $B_0 = S_0$. Let $t_m = m\Delta t$ for $m = 0, 1, \ldots, n$, and define a discrete time process $X = \{X_{t_m} = B_m\}$. Note that $\log X$ can be written in the following fashion:

$$\log X_{t_m} = \log S_0 + \sigma\sqrt{\Delta t}\sum_{i=1}^{m} Y_i$$

where Y_i are iid random variables with

$$\mathbb{P}(Y_i = +1) = p = \frac{e^{\mu\Delta t} - d}{u - d}, \quad \mathbb{P}(Y_i = -1) = 1 - p.$$

(a) Show that $\{\log X_{t_m} : m = 0, 1, \ldots, n\}$ has independent and stationary increments. That is,

$$\log X_{t_1} - \log X_{t_0}, \ldots, \log X_{t_n} - \log X_{t_{n-1}}$$

are iid random variables.

(b) Identify the limit of $E[\log X_{t_m}]$ and $\text{Var}[\log X_{t_m}]$ when $t_m \to t$.

(c) In light of the central limit theorem, assume that as $t_m \to t$, X_{t_m} converges to a normal distribution. Argue that the limit distribution is the same as the distribution of $\log S_t$.

(d) Explain intuitively that $\{X_{t_m}\}$ converges to $\{S_t\}$ on interval $[0, T]$.

MATLAB® Problems

3.A Write a function using the "function" command to price call and put options with the same strike price and maturity, via the binomial tree method. Save this function as a ".m" file. The function should include the following parameters as input:

u	=	factor by which the stock price moves up in each period,
d	=	factor by which the stock price moves down in each period,
N	=	number of periods,
S_0	=	initial stock price,
r	=	risk-free interest rate,
T	=	maturity,
K	=	strike price.

The output of the function should be the call option price and the put option price. Observe that in a binomial tree model, the possible values of the stock price at the N-th period are

$$S_0 u^i d^{N-i}, \quad i = N, N-1, ..., 0.$$

In the programming you may want to use matrices. There are various MAT-LAB commands to initialize a variable as a matrix. For example,

    "zeros(m,n)": returns an $m \times n$ matrix with each entry 0,

    "ones(m,n)": returns an $m \times n$ matrix with each entry 1,

    "eye(n)": returns an $n \times n$ identity matrix.

3.B Suppose that the stock price S is a geometric Brownian motion under the risk-neutral probability measure:

$$S_t = S_0 \exp\left\{ \left(r - \frac{1}{2}\sigma^2 \right) t + \sigma W_t \right\}$$

with the initial price $S_0 = 20$. Use the binomial tree approximation with $N = 30, 60, 120$ periods, respectively, to price the call and put options with strike price $K = 20$ and maturity $T = 1$ year, assuming that the risk-free interest rate is $r = 8\%$ annually and $\sigma = 0.3$. Compare your results with the theoretical Black–Scholes prices, namely, BLS_Call and BLS_Put.

Chapter 4

Monte Carlo Simulation

*— used to estimate integrals & expected values
to find expected values.*

Monte Carlo simulation is a very flexible and powerful tool for estimating integrals and expected values. Since most of the quantitative analysis in finance or risk management involves computing quantities that are indeed expected values, Monte Carlo simulation is widely used in the financial industry. This chapter aims to give a quick introduction to Monte Carlo simulation, as well as its pros and cons.

4.1 Basics of Monte Carlo Simulation

Consider the generic question of estimating the expected value of a function of some random variable X: $\mu = E(h(X))$ *mean ≡ expected value*

$$\mu = E[h(X)].$$

A plain Monte Carlo simulation scheme can be roughly divided into two steps:

1. Generate samples, or independent identically distributed (iid) random variables $X_1, X_2, \ldots, X_n$, that have the same distribution as X.

2. The estimate of the expected value μ is defined to be the sample average

$$\hat{\mu} = \frac{1}{n}[h(X_1) + h(X_2) + \cdots + h(X_n)].$$

Sometimes we simply refer to the samples $X_1, X_2, \ldots, X_n$ as *iid copies* of X. The number of samples n is the *sample size*, which is usually chosen to be a large number. It should be noted that μ, the quantity we wish to estimate, is an unknown *fixed* number, whereas the Monte Carlo estimate $\hat{\mu}$ is a *random variable*. The value of $\hat{\mu}$ will vary depending on the samples.

It is possible to design many different Monte Carlo simulation algorithms for estimating the same expected value μ. We briefly mention a couple of alternatives.

(a) *Importance Sampling:* Assuming that X admits a density f, we can write

$$\mu = \int_{\mathbb{R}} h(x)f(x)\,dx.$$

Consider an arbitrary nonzero density function $g(x)$. It follows that

$$\mu = \int_{\mathbb{R}} h(x)\frac{f(x)}{g(x)} \cdot g(x)\,dx = E\left[h(Y)\frac{f(Y)}{g(Y)}\right],$$

where Y is a random variable with density g. The corresponding Monte Carlo estimate for μ is

$$\hat{\mu} = \frac{1}{n}\sum_{i=1}^{n} h(Y_i)\frac{f(Y_i)}{g(Y_i)},$$

average of the Sum.

where $Y_1, Y_2, \ldots, Y_n$ are iid copies of Y.

(b) *Control Variates:* Suppose that there is a random variable Y such that $E[Y] = 0$. Then one can write

$$\mu = E[h(X) + Y].$$

The corresponding Monte Carlo estimate for μ is

$$\hat{\mu} = \frac{1}{n}\sum_{i=1}^{n}[h(X_i) + Y_i],$$

where $(X_1, Y_1), (X_2, Y_2), \ldots, (X_n, Y_n)$ are iid copies of (X, Y).

All these different Monte Carlo schemes can be generically described as follows. Let H be a random variable such that $\mu = E[H]$. Then the corresponding Monte Carlo estimate is given by

$$\hat{\mu} = \frac{1}{n}\sum_{i=1}^{n} H_i,$$

where $H_1, H_2, \ldots, H_n$ are iid copies of H.

The underlying principle for Monte Carlo simulation is the strong law of large numbers. That is, as the sample size n tends to infinity,

larger sample size,

$$\hat{\mu} = \frac{1}{n}(H_1 + H_2 + \cdots + H_n) \to E[H] = \mu$$

Closer to true mean.

with probability one. Therefore, the estimate $\hat{\mu}$ is expected to be close to the true value μ when n is large.

How?

Remark 4.1. Since a probability can be expressed as an expected value, Monte Carlo simulation is commonly used for estimating probabilities as well. For example, for any subset $A \subseteq \mathbb{R}$, one can write

$$\mathbb{P}(X \in A) = E[h(X)],$$

where h is an indicator function defined by

$$h(x) = \begin{cases} 1 & \text{if } x \in A, \\ 0 & \text{otherwise.} \end{cases}$$

In this case, the plain Monte Carlo estimate for $\mathbb{P}(X \in A)$ is just the proportion of samples that fall into set A.

4.2 Standard Error and Confidence Interval

Let μ be the unknown quantity of interest and H a random variable such that $\mu = E[H]$. A Monte Carlo estimate for μ is

$$\hat{\mu} = \frac{1}{n}\sum_{i=1}^{n} H_i,$$

where $H_1, H_2, \ldots, H_n$ are iid copies of H. As we have mentioned, the strong law of large numbers asserts that $\hat{\mu}$ is close to μ when n is large. But how close? Since $\hat{\mu}$ is a random variable, so is the error $\hat{\mu} - \mu$. Therefore, what we are really looking for is the distribution of this error, not the error bound in the usual sense.

Denote by σ_H^2 the variance of H. It follows from the central limit theorem that as $n \to \infty$,

$$\frac{H_1 + H_2 + \cdots + H_n - n\mu}{\sigma_H\sqrt{n}} = \frac{\sqrt{n}(\hat{\mu} - \mu)}{\sigma_H}$$

Central limit theorem?

converges to the standard normal distribution. That is, for every $a \in \mathbb{R}$,

$$\mathbb{P}\left\{\frac{\sqrt{n}(\hat{\mu} - \mu)}{\sigma_H} \leq a\right\} \rightarrow \Phi(a). \tag{4.1}$$

In other words, the error $\hat{\mu} - \mu$ is approximately normally distributed with mean 0 and variance σ_H^2/n. This asymptotic analysis also produces *confidence intervals* for the Monte Carlo estimate $\hat{\mu}$. More precisely, it follows from (4.1) that the $1 - \alpha$ confidence interval for μ is approximately

$$\hat{\mu} \pm z_{\alpha/2}\frac{\sigma_H}{\sqrt{n}},$$

where $z_{\alpha/2}$ is defined by the equation $\Phi(-z_{\alpha/2}) = \alpha/2$. Confidence intervals are *random* intervals. A $1 - \alpha$ confidence interval has probability $1 - \alpha$ of covering the true value μ. Note that the width of a confidence interval decreases as the sample size increases. If one quadruples the sample size, the width is reduced by half.

In practice, the standard deviation σ_H is rarely known. Instead, the sample standard deviation

$$s_H = \sqrt{\frac{1}{n-1}\sum_{i=1}^{n}(H_i - \hat{\mu})^2} = \sqrt{\frac{1}{n-1}\left(\sum_{i=1}^{n}H_i^2 - n\hat{\mu}^2\right)}$$

is used in place of σ_H. This substitution is appropriate since s_H converges to σ_H with probability one when the sample size n tends to infinity, and hence the central limit theorem still holds if σ_H is replaced by s_H. The empirical $1 - \alpha$ confidence interval thus becomes

$$\hat{\mu} \pm z_{\alpha/2}\frac{s_H}{\sqrt{n}}.$$

The quantity $s_H/\sqrt{n}$ is often said to be the *standard error* of $\hat{\mu}$. That is,

$$\text{S.E.} = \sqrt{\frac{1}{n(n-1)}\left(\sum_{i=1}^{n}H_i^2 - n\hat{\mu}^2\right)}.$$

Therefore, the commonly used 95% confidence interval is just the estimate $\hat{\mu}$ plus/minus twice the standard error (note that $z_{0.025} \approx 2$). In Monte Carlo simulation, it is customary to report both the estimate and the standard error.

Remark 4.2. The variance of H determines the width of a confidence interval and in some sense, the size of the error $\hat{\mu} - \mu$. Given a fixed sample size, the smaller the variance, the tighter the confidence interval, and hence the more accurate the estimate. This leads to the following criterion: *when comparing two estimates, the one with the smaller variance is more efficient.* Naturally, such a statement is a great simplification of the more scientific efficiency criteria that also take into consideration the computational effort of generating samples [11]. Nonetheless, it is a valuable guiding principle and will be used throughout the book to analyze the efficiency of various Monte Carlo schemes.

Remark 4.3. The Monte Carlo estimates that we have discussed so far are all *unbiased*, that is, *estimate (expected value) = true mean*

$$E[\hat{\mu}] = \mu.$$

Unbiasedness is a desirable property, but it is not always attainable. For example, consider estimating the price of a lookback call option with payoff

take advantage of widest differential between Strike price and Stock price.

$$\left(\max_{0 \leq t \leq T} S_t - K \right)^+$$

at maturity T. Except for some rare occasions, it is impossible to exactly simulate the maximum of a continuous time sample path. Instead, one often uses the discrete time analogue

$$\max_{i=1,\dots,m} S_{t_i}, \quad t_i = iT/m$$

as an approximation. The plain Monte Carlo estimate for the price is just the sample mean of iid copies of

$$e^{-rT} \left(\max_{i=1,\dots,m} S_{t_i} - K \right)^+$$

steps

under the risk-neutral probability measure. This estimate has a negative bias since

$$\max_{i=1,\dots,m} S_{t_i} \leq \max_{0 \leq t \leq T} S_t.$$

Note that the bias is very different from the random error of a Monte Carlo estimate. While the latter decreases when the sample size becomes larger, the bias can only be reduced by increasing the discretization parameter m.

bias linked to steps?

4.3   Examples of Monte Carlo Simulation

We first study a few simple examples to illustrate the basic structure and techniques of Monte Carlo simulation. For each example, we report not only the estimate but also the standard error because the latter indicates how accurate the former is.

Example 4.1. Simulate W_T. Consider the problem of estimating the price of a call option under the assumption that the underlying stock price is a geometric Brownian motion.

SOLUTION: Recall that the price of a call option with strike price K and maturity T is

$$price = v = E[e^{-rT}(S_T - K)^+],$$

where r is the risk-free interest rate. The expected value is taken under the risk-neutral probability measure, where the stock price is a geometric Brownian motion with drift r:

$$S_T = S_0 \exp\left\{ \left(r - \frac{1}{2}\sigma^2 \right) T + \sigma W_T \right\}.$$

In order to generate samples of the option payoff, it suffices to generate samples of W_T. Since W_T is normally distributed with mean 0 and variance T, one can write $W_T = \sqrt{T}Z$ for some standard normal random variable Z.

Pseudocode:

> for $i = 1, 2, \ldots, n$
>> generate Z_i from $N(0,1)$
>>
>> set $Y_i = S_0 \exp\left\{ \left(r - \frac{1}{2}\sigma^2 \right) T + \sigma\sqrt{T}Z_i \right\}$
>>
>> set $X_i = e^{-rT}(Y_i - K)^+$
>
> compute the estimate $\hat{v} = \frac{1}{n}(X_1 + X_2 + \cdots + X_n)$
>
> compute the standard error S.E. $= \sqrt{\dfrac{1}{n(n-1)}\left(\sum_{i=1}^{n} X_i^2 - n\hat{v}^2 \right)}.$

The simulation results are reported in Table 4.1 with the parameters given by

$$S_0 = 50, \quad r = 0.05, \quad \sigma = 0.2, \quad T = 1.$$

For comparison, the theoretical values are calculated from the Black–Scholes formula BLS_Call; see Example 2.1.

Table 4.1: Monte Carlo simulation for call options

	Sample size $n = 2500$			Sample size $n = 10000$		
Strike price K	40	50	60	40	50	60
Theoretical value	12.2944	5.2253	1.6237	12.2944	5.2253	1.6237
M.C. Estimate	12.2677	5.2992	1.6355	12.3953	5.2018	1.6535
S.E.	0.1918	0.1468	0.0873	0.0964	0.0727	0.0438

Note that the standard errors of the estimates drop roughly 50% when the sample size is quadrupled. ∎

Example 4.2. Simulate a Brownian Motion Sample Path. Consider a discretely monitored average price call option whose payoff at maturity T is

$$\left(\frac{1}{m}\sum_{i=1}^{m} S_{t_i} - K\right)^+,$$

where $0 < t_1 < \cdots < t_m = T$ are a fixed set of dates. Assume that under the risk-neutral probability measure,

$$S_t = S_0 \exp\left\{\left(r - \frac{1}{2}\sigma^2\right)t + \sigma W_t\right\}.$$

Estimate the price of the option.

SOLUTION: The key issue is to generate iid copies of the discrete time sample path $(S_{t_1}, S_{t_2}, \ldots, S_{t_m})$. They should be simulated sequentially since

$$S_{t_{i+1}} = S_{t_i} \exp\left\{\left(r - \frac{1}{2}\sigma^2\right)(t_{i+1} - t_i) + \sigma(W_{t_{i+1}} - W_{t_i})\right\},$$

and $(W_{t_1} - W_{t_0}, \ldots, W_{t_m} - W_{t_{m-1}})$ are *independent* normal random variables. Below is the pseudocode for generating *one* sample path.

Pseudocode:

for $i = 1, \ldots, m$

generate Z_i from $N(0, 1)$

set $S_{t_i} = S_{t_{i-1}} \exp\left\{\left(r - \frac{1}{2}\sigma^2\right)(t_i - t_{i-1}) + \sigma\sqrt{t_i - t_{i-1}}Z_i\right\}$

compute the discounted payoff $X = e^{-rT}\left(\frac{1}{m}\sum_{i=1}^{m} S_{t_i} - K\right)^+.$

The Monte Carlo algorithm will repeat the above steps n times to obtain n sample paths and n iid copies of X, say $X_1, \ldots, X_n$. The estimate and its standard error are given by

— random variable

$$\hat{v} = \frac{1}{n}(X_1 + X_2 + \cdots + X_n),$$

$$\text{S.E.} = \sqrt{\frac{1}{n(n-1)} \left(\sum_{k=1}^{n} X_k^2 - n\hat{v}^2 \right)}.$$

The simulation results are reported in Table 4.2. The parameters are set to be

$$S_0 = 50, \quad r = 0.05, \quad \sigma = 0.2, \quad T = 1, \quad m = 12, \quad t_i = \frac{i}{12}.$$

Explicit formula for the option price is not available.

Table 4.2: Monte Carlo simulation for average price call options

	Sample size $n = 2500$			Sample size $n = 10000$		
Strike price K	40	50	60	40	50	60
M.C. Estimate	10.7487	3.0730	0.3837	10.8810	3.0697	0.3490
S.E.	0.1183	0.0846	0.0332	0.0597	0.0422	0.0152

As in the previous example, the standard errors of the estimates reduce roughly by half when the sample size increases fourfold. ∎

→ To check if the code is working well.

Example 4.3. Simulate 2D Jointly Normal Random Vectors. Estimate the price of a spread call option whose payoff at maturity T is

$$(X_T - Y_T - K)^+,$$

where $\{X_t\}$ and $\{Y_t\}$ are the prices of two underlying assets. Assume that under the risk-neutral probability measure,

$$X_t = X_0 \exp\left\{ \left(r - \frac{1}{2}\sigma_1^2 \right) t + \sigma_1 W_t \right\},$$

$$Y_t = Y_0 \exp\left\{ \left(r - \frac{1}{2}\sigma_2^2 \right) t + \sigma_2 B_t \right\},$$

where (W, B) is a two-dimensional Brownian motion with covariance matrix

$$\Sigma = \begin{bmatrix} 1 & \rho \\ \rho & 1 \end{bmatrix}.$$

SOLUTION: By assumption (W_T, B_T) is a jointly normal random vector with mean 0 and covariance matrix

$$T \cdot \begin{bmatrix} 1 & \rho \\ \rho & 1 \end{bmatrix}.$$

If the two Brownian motions W and B are uncorrelated (that is, if $\rho = 0$), then the simulation is straightforward—one could just sample two independent standard normal random variables Z_1 and Z_2, and let

$$X_T = X_0 \exp\left\{ \left(r - \frac{1}{2}\sigma_1^2 \right) T + \sigma_1 \sqrt{T} Z_1 \right\},$$

$$Y_T = Y_0 \exp\left\{ \left(r - \frac{1}{2}\sigma_2^2 \right) T + \sigma_2 \sqrt{T} Z_2 \right\}.$$

When $\rho \neq 0$, the simulation is more involved. Recall that if Z_1 and Z_2 are two independent standard normal random variables, then

$$Z = \begin{bmatrix} Z_1 \\ Z_2 \end{bmatrix}$$

is a two-dimensional standard normal random vector (see Appendix A). Therefore, for any 2×2 matrix $C = [C_{ij}]$, the random vector

$$R = CZ = \begin{bmatrix} C_{11}Z_1 + C_{12}Z_2 \\ C_{21}Z_1 + C_{22}Z_2 \end{bmatrix}$$

is jointly normal with mean 0 and covariance matrix CC'. If there exists a matrix C such that

$$CC' = \begin{bmatrix} 1 & \rho \\ \rho & 1 \end{bmatrix}, \tag{4.2}$$

then $\sqrt{T}R$ will have the same distribution as (W_T, B_T) and we can let

$$X_T = X_0 \exp\left\{ \left(r - \frac{1}{2}\sigma_1^2 \right) T + \sigma_1 \sqrt{T} R_1 \right\},$$

$$Y_T = Y_0 \exp\left\{ \left(r - \frac{1}{2}\sigma_2^2 \right) T + \sigma_2 \sqrt{T} R_2 \right\}.$$

There are many choices of C that satisfy (4.2). A particularly convenient one is when C is a lower triangular matrix:

$$C = \begin{bmatrix} C_{11} & 0 \\ C_{21} & C_{22} \end{bmatrix}.$$

In this case

$$CC' = \begin{bmatrix} C_{11}^2 & C_{11}C_{21} \\ C_{21}C_{11} & C_{21}^2 + C_{22}^2 \end{bmatrix} = \begin{bmatrix} 1 & \rho \\ \rho & 1 \end{bmatrix}.$$

Taking $C_{11} = 1$, we arrive at

$$C = \begin{bmatrix} 1 & 0 \\ \rho & \sqrt{1-\rho^2} \end{bmatrix}.$$

Below is the pseudocode for estimating the price of the spread call option.

Pseudocode:

> set $C_{11} = 1$, $C_{21} = \rho$, $C_{22} = \sqrt{1-\rho^2}$
> for $i = 1, 2, \ldots, n$
>> generate Z_1 and Z_2 from $N(0,1)$
>> set $R_1 = C_{11}Z_1$ and $R_2 = C_{21}Z_1 + C_{22}Z_2$
>> set $X_i = X_0 \exp\left\{ \left(r - \frac{1}{2}\sigma_1^2\right) T + \sigma_1\sqrt{T}R_1 \right\}$
>> set $Y_i = Y_0 \exp\left\{ \left(r - \frac{1}{2}\sigma_2^2\right) T + \sigma_2\sqrt{T}R_2 \right\}$
>> compute the discounted payoff $H_i = e^{-rT}(X_i - Y_i - K)^+$
>> compute the estimate $\hat{v} = \frac{1}{n}(H_1 + H_2 + \cdots + H_n)$
>> compute the standard error S.E. $= \sqrt{\frac{1}{n(n-1)} \left(\sum_{i=1}^{n} H_i^2 - n\hat{v}^2 \right)}$.

The simulation results are reported in Table 4.3. The parameters are given by

$$X_0 = 50, \quad Y_0 = 45, \quad r = 0.05, \quad \sigma_1 = 0.2, \quad \sigma_2 = 0.3, \quad \rho = 0.5, \quad T = 1.$$

Table 4.3: Monte Carlo simulation for spread call options

	Sample size $n = 2500$			Sample size $n = 10000$		
Strike price K	0	5	10	0	5	10
M.C. Estimate	7.9593	4.9831	2.7024	7.9019	5.0056	2.8861
S.E.	0.1680	0.1330	0.0990	0.0838	0.0683	0.0521

The matrix C is a special case of the *Cholesky factorization*. It can be generalized to simulate higher dimensional jointly normal random vectors. We will take the discussion further in Chapter 5. ∎

We should give a couple of examples to demonstrate that neither does every Monte Carlo estimate take the form of sample average, nor is the construction of an efficient Monte Carlo scheme always automatic. The first example is concerned with estimating value-at-risk, which is essentially a quantile of an unknown distribution. The difficulty there is the construction of confidence intervals. The second example is about estimating the probability of a large loss in a credit risk model. Such probabilities are usual very small, which renders the plain Monte Carlo scheme quite inefficient or even infeasible.

Example 4.4. Simulate Value-at-Risk. Denote by X_i the daily return of a portfolio. Assume that $X = \{X_1, X_2, \ldots\}$ is a *Markov chain*, that is, the conditional distribution of X_{i+1} given $(X_i, X_{i-1}, \ldots, X_1)$ only depends on X_i for each i. Let X be autoregressive conditional heteroskedastic (ARCH) in the sense that given $X_i = x$, X_{i+1} is normally distributed with mean zero and variance $\beta_0 + \beta_1 x^2$ for some $\beta_0 > 0$ and $0 < \beta_1 < 1$. The total return within an m-day period is

$$S = \sum_{i=1}^{m} X_i.$$

Assuming that X_1 is a standard normal random variable, estimate the value-at-risk at the confidence level $1 - p$.

SOLUTION: The simulation of a Markov chain is done sequentially as the distribution of X_{i+1} depends on the value of X_i. Below is the pseudocode for generating one sample of S:

Pseudocode for one sample of the total return S:

generate X_1 from $N(0,1)$
for $i = 2, 3, \ldots, m$
    generate Z from $N(0,1)$
    set $X_i = \sqrt{\beta_0 + \beta_1 X_{i-1}^2} \cdot Z$
set $S = \sum_{i=1}^{m} X_i.$

The Monte Carlo scheme will repeat the above steps n times to generate n iid copies of S, say $S_1, \ldots, S_n$. Recall that the value-at-risk (VaR) at the confidence level $1 - p$ is defined by

$$\mathbb{P}(S \leq -\text{VaR}) = p.$$

Thus it makes sense to estimate VaR by a number $\hat{x}$ such that the fraction of samples at or below level $-\hat{x}$, i.e.,

$$\frac{\text{number of samples among } \{S_1, \ldots, S_n\} \text{ at or below } -\hat{x}}{n},$$

is close to p. Consider the *order statistics* $\{S_{(1)}, \ldots, S_{(n)}\}$, which is a permutation of $\{S_1, \ldots, S_n\}$ in an increasing order:

$$S_{(1)} \leq S_{(2)} \leq \cdots \leq S_{(n)}.$$

A common choice for $\hat{x}$ is to let $k = [np]$ (the integer part of np) and set

$$-\hat{x} = S_{(k)}.$$

Note that $\hat{x}$ is neither the sample average nor an unbiased estimate.

The difficulty here lies in the construction of confidence intervals for this estimate. Even though there are various approaches, we will only describe a simple method based on the order statistics, which works best when the sample size n is large. Suppose that one is interested in a $1 - \alpha$ confidence interval. The goal is to find integers $k_1 < k_2$ such that

$$\mathbb{P}(S_{(k_1)} \leq -\text{VaR} < S_{(k_2)}) = 1 - \alpha. \tag{4.3}$$

If this can be done exactly or approximately, then $(-S_{(k_2)}, -S_{(k_1)}]$ serves as a $1 - \alpha$ confidence interval. To this end, observe that the left-hand-side of (4.3) is just $\mathbb{P}(k_1 \leq Y < k_2)$, where Y denotes the number of samples that are less than or equal to $-\text{VaR}$. Since Y is binomial with parameters (n, p), this probability equals

$$\sum_{j=k_1}^{k_2-1} \binom{n}{j} (1 - p)^j p^{n-j}.$$

When n is large, a direct evaluation of this summation is difficult. However, one can use the normal approximation of binomial distributions (see Exercise 4.1) to conclude that

$$\mathbb{P}(S_{(k_1)} \leq -\text{VaR} < S_{(k_2)}) \approx 1 - \Phi\left(\frac{k_1 - np}{\sqrt{np(1-p)}}\right) - \Phi\left(-\frac{k_2 - np}{\sqrt{np(1-p)}}\right),$$

which leads to the following choice of k_1 and k_2 (taking integer part if necessary):

$$k_1 = np - \sqrt{np(1-p)}z_{\alpha/2}, \quad k_2 = np + \sqrt{np(1-p)}z_{\alpha/2}.$$

The simulation results are reported in Table 4.4, given $\beta_0 = \beta_1 = 0.5$ and $m = 10$. We estimate the value-at-risk at the confidence level $1 - p$ and give a 95% confidence interval for $p = 0.05, 0.02, 0.01$, respectively. The sample size is $n = 10000$.

Table 4.4: Monte Carlo simulation for value-at-risk

p	0.05	0.02	0.01
k	500	200	100
(k_1, k_2)	(458,542)	(173, 227)	(80,120)
Estimate $\hat{x}$	4.9978	6.4978	7.7428
95% C.I.	$[4.8727, 5.1618]$	$[6.3381, 6.6835]$	$[7.4650, 8.0430]$

The probability in (4.3) does not depend on the underlying distribution of S. For this reason, the confidence intervals constructed here are said to be *distribution free* or *nonparametric*. ∎

Example 4.5. Difficulty in Estimating Small Probabilities. Consider the one-factor portfolio credit risk model in Example 1.9. Let c_k denote the loss from the default of the k-th obligor. Then the total loss is

$$L = \sum_{k=1}^{m} c_k 1_{\{X_k \geq x_k\}}.$$

Estimate the probability that L exceeds a given large threshold h.

SOLUTION: The simulation algorithm is straightforward.

Pseudocode:

> for $i = 1, 2, \ldots, n$
>> generate independent samples $Z, \varepsilon_1, \ldots, \varepsilon_m$ from $N(0,1)$
>> compute $X_k = \rho_k Z + \sqrt{1 - \rho_k^2} \varepsilon_k$ for $k = 1, \ldots, m$
>> compute $L = \sum_{k=1}^{m} c_k 1_{\{X_k \geq x_k\}}$
>> set $H_i = 1$ if $L > h$; set $H_i = 0$ otherwise
>> compute the estimate $\hat{v} = \frac{1}{n}(H_1 + H_2 + \cdots + H_n)$
>> compute the standard error S.E. $= \sqrt{\frac{1}{n(n-1)} \left(\sum_{i=1}^{n} H_i^2 - n\hat{v}^2 \right)}.$

Set $m = 3$, $c_1 = 2$, $c_2 = 1$, $c_3 = 4$, $\rho_1 = 0.2$, $\rho = 0.5$, $\rho_3 = 0.8$. The levels are assumed to be $x_1 = 1$, $x_2 = 1$, $x_3 = 2$. The simulation results are reported in Table 4.5, where we have also included

$$Empirical\ Relative\ Error = \frac{Standard\ Error}{Estimate}.$$

Table 4.5: Monte Carlo simulation for a credit risk model

	Sample size $n = 2500$			Sample size $n = 10000$		
Threshold h	1	2	4	1	2	4
M.C. Estimate	0.1840	0.0476	0.0136	0.1780	0.0528	0.0150
S.E.	0.0078	0.0043	0.0023	0.0038	0.0022	0.0012
R.E.	4.24%	9.03%	16.91%	2.13%	4.17%	8.00%

	Sample size $n = 10000$			Sample size $n = 40000$		
Threshold h	6	8	10	6	8	10
M.C. Estimate	0.0028	0.0000	0.0000	0.0032	0.0000	0.0000
S.E.	0.0005	0.0000	0.0000	0.0003	0.0000	0.0000
R.E.	18.87%	NaN	NaN	8.76%	NaN	NaN

An interesting observation is that as the threshold h gets larger, the probability gets smaller and the quality of the estimates deteriorates. The reason is that only for a very small fraction of samples will the total loss exceed the large threshold h. With so few hits, the estimate cannot be accurate. Therefore, the plain Monte Carlo is inefficient for estimating small probabilities; see also Exercise 4.2. Clearly, a more efficient Monte Carlo scheme is needed for estimating such small quantities. ∎

4.4 Summary

Monte Carlo simulation is a very useful tool for the quantitative analysis of financial models. It is well suited for parallel computing, and its flexibility can accommodate complicated models that are otherwise inaccessible.

Monte Carlo simulation is a random algorithm. A different run of simulation will yield a different estimate. It is very different from those deterministic numerical schemes for evaluating integrals, which are usually designed for problems of low dimensions. Since many of the pricing problems in financial engineering are intrinsically problems of evaluating integrals of large or infinite dimensions, these deterministic algorithms are not well suited for this type of tasks. On the contrary, the central limit theorem asserts that the standard error of a Monte Carlo estimate decays in the order of $O(1/\sqrt{n})$ with respect to the sample size n, *regardless of the dimension.*

But the Monte Carlo method is not without shortcomings. Even though it is often possible to improve the efficiency of a given Monte Carlo scheme, little can be done to accelerate the convergence above the rate $O(1/\sqrt{n})$. A large sample size is often required in order to achieve a desirable accuracy level.

Finally, the design of Monte Carlo schemes is not as straightforward as one might think. This is especially true when the quantity of interest is associated with events of small probabilities, which is a common scenario in risk analysis. Here one has to be very cautious, since it is not uncommon that a seemingly very accurate estimate (i.e., an estimate with a very small standard error; see Exercise 4.F, for example) can be far off from the true value. Theoretical justification of a Monte Carlo scheme should be provided whenever possible.

Exercises

Pen-and-Paper Problems

4.1 Suppose that $X_1, \ldots, X_n$ are iid Bernoulli random variables with parameter p. Then

$$S_n = X_1 + \ldots + X_n$$

is a binomial random variable with parameters (n, p). Use the central limit theorem to explain that, when n is large, the binomial distribution with parameters (n, p) can be approximated by the normal distribution with mean np and variance $np(1 - p)$. That is, for $x \in \mathbb{R}$,

$$\mathbb{P}\left(\frac{S_n - np}{\sqrt{np(1 - p)}} \leq x \right) \approx \Phi(x).$$

4.2 Suppose that $X_1, \ldots, X_n$ are iid Bernoulli random variables with unknown parameter p. The average

$$\bar{X} = \frac{1}{n}(X_1 + \cdots + X_n)$$

can be used to estimate p.

(a) What are the expected value and standard deviation of $\bar{X}$?

(b) Write down a 95% confidence interval for p.

(c) Define the relative error as

$$Relative\ Error = \frac{\text{Standard Deviation of } \bar{X}}{\text{Expected Value of } \bar{X}}.$$

How large should n be so that the relative error is at most 5%?

4.3 In a class of 100 students, each student is asked to run a simulation to estimate the price of an option and provide a 95% confidence interval, independently from others. What is the distribution of the number of confidence intervals that cover the true value of the option price? Is it likely that all the confidence intervals cover the true value of the option price?

4.4 Let X be a random variable with density $f(x)$. It is easy to see that estimating the integral

$$\int_{\mathbb{R}} h(x)f(x)\, dx$$

amounts to estimating the expected value $E[h(X)]$. Use this observation to design Monte Carlo schemes for estimating the following integrals:

$$\int_0^\infty e^{-x} \sin(x)\, dx, \quad 4\int_0^1 \sqrt{1 - x^2}\, dx, \quad \int_0^\infty \frac{1}{\sqrt{x}} e^{-x^2}\, dx.$$

Write down the pseudocode (it should report both the estimate and the standard error).

4.5 Let $\hat{\mu}$ be an estimate for some unknown quantity μ. The difference $E[\hat{\mu}] - \mu$ is said to be the *bias* of $\hat{\mu}$. The *mean square error* (M.S.E.) of $\hat{\mu}$ is defined to be $E[(\hat{\mu} - \mu)^2]$. Show that

$$\text{M.S.E.} = (\text{Bias of } \hat{\mu})^2 + \text{Var}[\hat{\mu}].$$

In general, it is beneficial to allocate the computational budget in order to balance bias and variance. The rule of thumb is to make the bias and the standard deviation of the estimate roughly the same order [11].

4.6 Samples of a random vector (X, Y) can often be drawn in a sequential manner: one first samples X from its marginal distribution and then samples Y from its conditional distribution given X. Explain that it is essentially what has been done in Example 4.3 to simulate the jointly normal random vector $R = (R_1, R_2)$ with mean 0 and covariance matrix Σ.

4.7 Let $W = \{W_t : t \geq 0\}$ be a standard Brownian motion. Consider the random vector (W_t, M_t), where M_t is the running maximum of W by time t, that is,

$$M_t = \max_{0 \leq s \leq t} W_s.$$

Show that the conditional distribution of M_T given $W_T = x$ is identical to the distribution of

$$\frac{1}{2}\left(x + \sqrt{x^2 + 2TY}\right),$$

where Y is an exponential random variable with rate one. Use this result to design a scheme to draw samples from

(a) (W_T, M_T);

(b) $(W_{t_1}, M_{t_1}, W_{t_2}, M_{t_2}, \ldots, W_{t_m}, M_{t_m})$ given $0 < t_1 < t_2 < \cdots < t_m = T$;

(c) (B_T, M_T^B), where $B_t = W_t + \theta t$ is Brownian motion with drift θ and M^B is the running maximum of B, that is,

$$M_t^B = \max_{0 \leq s \leq t} B_s;$$

(d) $(B_{t_1}, M_{t_1}^B, B_{t_2}, M_{t_2}^B, \ldots, B_{t_m}, M_{t_m}^B)$ given $0 < t_1 < t_2 < \cdots < t_m = T$.

Write down the pseudocode. *Hint:* Recall Exercise 2.12 for (c) and (d).

4.8 Consider the following two Monte Carlo schemes for estimating $\mu = E[h(X) + f(X)]$. The total sample size is $2n$ in both schemes.

(a) *Scheme I: Use Same Random Numbers.* Generate $2n$ iid copies of X, say $\{X_1, \ldots, X_{2n}\}$. The estimate is

$$\hat{\mu}_1 = \frac{1}{2n} \sum_{i=1}^{2n} [h(X_i) + f(X_i)].$$

(b) *Scheme II: Use Different Random Numbers.* Write $\mu = E[h(X)] + E[f(X)]$ and estimate the two expected values separately. Generate $2n$ iid copies of X, say $\{X_1, \dots, X_n, Y_1, \dots, Y_n\}$. The estimate is

$$\hat{\mu}_2 = \frac{1}{n} \sum_{i=1}^{n} h(X_i) + \frac{1}{n} \sum_{i=1}^{n} f(Y_i).$$

Show that both $\hat{\mu}_1$ and $\hat{\mu}_2$ are unbiased estimates for μ, but $\hat{\mu}_1$ always has a smaller variance because

$$\text{Var}[\hat{\mu}_1] - \text{Var}[\hat{\mu}_2] = -\frac{1}{2n} \text{Var}[h(X) - f(X)] \leq 0.$$

This exercise shows that if one wants to estimate the price of a financial instrument such as straddle (the combination of a call and a put with the same strike price), it is beneficial to use the same random numbers to estimate its price altogether rather than estimate the call and the put prices separately.

MATLAB® Problems

In Exercises 4.A – 4.C, assume that the underlying stock price is a geometric Brownian motion under the risk-neutral probability measure:

$$S_t = S_0 \exp \left\{ \left(r - \frac{1}{2}\sigma^2 \right) t + \sigma W_t \right\},$$

where W is a standard Brownian motion and r is the risk-free interest rate.

4.A Write a function to estimate the price of a binary put option with maturity T and payoff

$$X = 1_{\{S_T \leq K\}}.$$

The input parameters are S_0, r, σ, K, T, and the sample size n. The function should output the estimate of the price and its standard error. Report your results for

$$S_0 = 30, \ r = 0.05, \ \sigma = 0.2, \ K = 30, \ T = 0.5, \ n = 10000.$$

Compare with the theoretical value of the option price.

4.B Write a function to estimate the price of a discretely monitored down-and-out call option with maturity T and payoff

$$(S_T - K)^+ \cdot 1_{\{\min(S_{t_1}, S_{t_2}, \dots, S_{t_m}) \geq b\}}.$$

The monitoring dates $t_1, \dots, t_m$ are prespecified and $0 < t_1 < \cdots < t_m = T$. The function should have the following parameters as input

$$S_0, \ r, \ \sigma, \ T, \ K, \ b, \ m, \ (t_1, \dots, t_m), \ n,$$

where n denotes the sample size. The output of the function should include the estimate for the option price and the standard error. Report your results for

$$S_0 = 50, \ r = 0.10, \ \sigma = 0.2, \ T = 1, \ K = 50, \ b = 45,$$
$$m = 12, \ t_i = iT/m, \ n = 10000.$$

4.C Consider a lookback put option with floating strike price and maturity T, whose payoff is

$$X = \max_{0 \le t \le T} S_t - S_T.$$

(a) Write a function to estimate the price of this lookback option, where the maximum of the stock price is approximated by

$$\max(S_{t_0}, S_{t_1}, \ldots, S_{t_m})$$

for some $0 = t_0 < t_1 < \cdots < t_m = T$. Let the input of the function be

$$S_0, \ r, \ \sigma, \ T, \ m, \ (t_0, t_1, \ldots, t_m), \ n,$$

where n is the sample size. Is the estimate unbiased? If it is not, is the bias positive or negative?

(b) Write a function that yields an unbiased estimate for the price of the lookback put option. The input of the function should be S_0, r, T, σ, n. *Hint:* Use Exercise 4.7 (a) and Lemma 2.2, or use Exercise 4.7 (c).

(c) Report your estimates and standard errors from (a) and (b) with the parameters given by

$$S_0 = 20, \ r = 0.03, \ \sigma = 0.2, \ T = 0.5, \ n = 10000.$$

For part (a) let $t_i = iT/m$ and $m = 10, 100, 1000$, respectively.

4.D Suppose that the stock price S is a geometric Brownian motion with jumps:

$$S_t = S_0 \exp\left\{ \left(\mu - \frac{1}{2}\sigma^2 \right) t + \sigma W_t + \sum_{i=1}^{N_t} Y_i \right\},$$

where W is a standard Brownian motion, $N = \{N_t : t \ge 0\}$ is a Poisson process with rate λ, and Y_i's are iid normal random variables with distribution $N(0, \nu^2)$. Assume that $W, N, \{Y_i\}$ are independent. Write a function to estimate the probability

$$\mathbb{P}\left(\max_{1 \le i \le m} S_{t_i} \ge b \right),$$

where $0 < t_1 < \cdots < t_m = T$ are prespecified dates and b is a given threshold. The function should have input parameters

$$S_0, \ \mu, \ \sigma, \ \lambda, \ \nu, \ b, \ T, \ m, \ (t_1, \ldots, t_m), \ n,$$

where n is the sample size. Report your estimate and its standard error for

$$S_0 = 50, \ \mu = 0.1, \ \sigma = 0.2, \ \lambda = 2, \ \nu = 0.3, \ b = 55, \ T = 1,$$

$$m = 50, \ t_i = iT/m, \ n = 10000.$$

4.E The setup is analogous to Example 4.4. Consider the following GARCH model: for every $i \geq 1$, $X_{i+1} = \sigma_{i+1} Z_{i+1}$, where Z_{i+1} is a standard normal random variable independent of $\{X_1, \ldots, X_i\}$ and

$$\sigma_{i+1}^2 = \beta_0 + \beta_1 X_i^2 + \beta_2 \sigma_i^2.$$

Assuming that X_1 is normally distributed as $N(0, \sigma_1^2)$ for some constant σ_1, write a function to estimate the value-at-risk of the total return

$$S = \sum_{i=1}^m X_i$$

at the confidence level $1 - p$. The input parameters of the function should be

$$\sigma_1, \ \beta_0, \ \beta_1, \ \beta_2, \ m, \ p, \ n,$$

where n denotes the sample size. Report your estimates and 95% confidence intervals for

$$\sigma_1 = 1, \ \beta_0 = 0.5, \ \beta_1 = 0.3, \ \beta_2 = 0.5, \ m = 10, \ n = 10000,$$

and $p = 0.05, 0.01$, respectively.

4.F Consider the problem of estimating $E[\exp\{\theta Z - \theta^2/2\}]$ for some constant θ and standard normal random variable Z. Use the plain Monte Carlo scheme with sample size $n = 1,000,000$. Report your simulation results for $\theta = 6$ and $\theta = 7$, respectively. Explain why your results are inconsistent with the theoretical value, which is one (see Exercise 1.7). *Hint:* The expected value can be written as

$$\int_{\mathbb{R}} e^{\theta x - \frac{1}{2}\theta^2} \frac{1}{\sqrt{2\pi}} e^{-\frac{1}{2}x^2} \, dx.$$

For what range of x does the majority of contribution to the integral come from? Can you trust the standard errors?

Inverse -Transform Method | techniques for generating
Acceptance rejection method | random variables
from a specialized
distributions

Chapter 5

Generating Random Variables

The success of a Monte Carlo scheme depends on its ability to repeatedly sample from a given distribution. At the heart of such sampling procedures is a mechanism for generating iid sequences of random numbers that are uniformly distributed on $[0, 1]$. In reality, such "random" numbers are not truly random. They are part of a very long sequence of numbers that are generated by completely deterministic computer algorithms to mimic the behavior of genuine random numbers. For this reason, such numbers are said to be *pseudo-random*. For all practical purposes, one can treat this sequence as if it were truly random. In MATLAB®, the function "rand" is used for generating such uniform random numbers.

Even though MATLAB has a large built-in collection of algorithms to generate samples from commonly used probability distributions, occasionally one would like to sample from some very specialized distributions. In this chapter, we will discuss some commonly used techniques for generating random variables or random vectors: the inverse transform method, the acceptance-rejection method, and the Cholesky factorization method for multivariate normal distributions.

5.1   Inverse Transform Method

The inverse transform method is by far the simplest and most commonly used method for generating random variables. The key observation is that any random variable can be represented as a function of a uniformly distributed random variable.

Consider the problem of generating samples from a probability distribution whose cumulative distribution function F is known. Define the *in-*

verse of F to be

$$F^{-1}(u) = \min\{x : F(x) \geq u\}$$

for every $u \in [0, 1]$. See Figure 5.1.

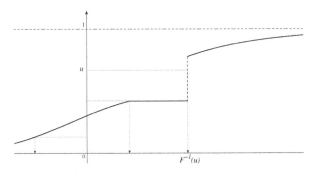

Figure 5.1: Inverse transform method.

Theorem 5.1. *If U is uniform on $[0, 1]$, then the cumulative distribution function of $X = F^{-1}(U)$ is F.*

PROOF. Observe that for any $x \in \mathbb{R}$,

$$\{U < F(x)\} \subseteq \{X \leq x\} \subseteq \{U \leq F(x)\}.$$

It follows that

$$F(x) = \mathbb{P}\{U < F(x)\} \leq \mathbb{P}(X \leq x) \leq \mathbb{P}(U \leq F(x)) = F(x),$$

or

$$\mathbb{P}(X \leq x) = F(x).$$

This completes the proof.   ∎

The inverse transform method is based on the preceding theorem. Once one obtains the inverse of F, then $X = F^{-1}(U)$ will have the desired distribution if U is uniformly distributed on $[0, 1]$.

Pseudocode for the inverse transform method:

generate a sample U from the uniform distribution on $[0, 1]$
set $X = F^{-1}(U)$.

Example 5.1. A Pareto distribution with parameters (a, b) has the density

$$f(x) = \begin{cases} ab^a/x^{a+1} & \text{if } x \geq b, \\ 0 & \text{otherwise.} \end{cases}$$

Here a and b are both positive constants. Determine F^{-1}.

SOLUTION: It is straightforward to compute the cumulative distribution function for the Pareto distribution from its density:

$$F(x) = \begin{cases} 1 - (b/x)^a & \text{if } x \geq b, \\ 0 & \text{otherwise.} \end{cases}$$

Therefore,

$$F^{-1}(u) = \frac{b}{(1-u)^{1/a}}$$

for all $u \in [0, 1]$. ∎

Example 5.2. Consider a discrete random variable X with distribution

$$\mathbb{P}(X = x_i) = p_i,$$

where $x_i \in \mathbb{R}$ and $p_i > 0$ are given such that

$$x_1 < x_2 < \cdots < x_m, \quad \sum_{i=1}^{m} p_i = 1.$$

Determine the inverse of its cumulative distribution function.

SOLUTION: Define $q_0 = 0$ and $q_k = p_1 + \cdots + p_k$ for $k = 1, \ldots, m$. The cumulative distribution function of X is

$$F(x) = q_j, \quad j = \max\{1 \leq i \leq m : x \geq x_i\},$$

with the convention that $\max\{\emptyset\} = 0$. It is straightforward to verify that

$$F^{-1}(u) = x_i, \quad \text{if } q_{i-1} < u \leq q_i$$

for $u \in [0, 1]$. ∎

5.2   Acceptance-Rejection Method

The acceptance-rejection method is another commonly used technique for generating random variables. Unlike the inverse transform method, it is not restricted to the univariate probability distributions.

For illustration, suppose that we are interested in generating samples from a target probability distribution with density f. Let g be an *alternative* density function, from which we know how to generate samples. Furthermore, assume that there exists a constant c such that

$$\frac{f(x)}{g(x)} \leq c \tag{5.1}$$

for all x. The algorithm is as follows. Generate a trial sample, say Y, from the density g. This sample will be accepted with probability

$$\frac{1}{c}\frac{f(Y)}{g(Y)}.$$

A trial sample that is not accepted will be discarded. Repeat this procedure until the desired sample size is reached.

Pseudocode for the acceptance-rejection method:

$(*)$ generate a trial sample Y from the density g

generate a sample U from the uniform distribution on $[0,1]$

accept Y if

$$U \leq \frac{1}{c}\frac{f(Y)}{g(Y)};$$

otherwise, discard Y and go to step $(*)$.

It is not difficult to explain why the acceptance-rejection method produces samples from the density f. Let X be a sample from the algorithm. Then

$$\mathbb{P}(X \in A) = \mathbb{P}(Y \in A \mid \text{sample } Y \text{ is accepted}).$$

By the law of total probability in Theorem 1.11,

$$\mathbb{P}(Y \text{ is accepted}) = \int \frac{1}{c}\frac{f(y)}{g(y)}g(y)dy = \frac{1}{c}$$

$$\mathbb{P}(Y \in A, \ Y \text{ is accepted}) = \int_A \frac{1}{c}\frac{f(y)}{g(y)}g(y)\,dy = \frac{1}{c}\int_A f(y)\,dy.$$

It follows that

$$\mathbb{P}(X \in A) = \int_A f(y)\, dy,$$

and hence the density of X is f. This calculation also shows that the overall probability of accepting a trial sample is $1/c$, or on average c samples from g are needed to generate one sample from f. Therefore, it is preferable that the constant c be close to 1 so that only a small fraction of samples from g will be rejected or wasted.

Example 5.3. Consider a bounded univariate density function f that is zero outside some interval $[a, b]$. It is very easy to design an acceptance-rejection algorithm that uses the uniform distribution on $[a, b]$ as the alternative sampling distribution. That is, $g(x) = 1/(b-a)$ for all $a \le x \le b$. The smallest, hence the optimal, constant c that satisfies the requirement (5.1) is

$$c = \max_{x \in [a,b]} \frac{f(x)}{g(x)} = (b-a) \max_{x \in [a,b]} f(x).$$

Carry this idea over to higher dimensions and construct an acceptance-rejection algorithm to generate samples that are uniformly distributed on the unit disc

$$\left\{ (x, y) : x^2 + y^2 \le 1 \right\}.$$

SOLUTION: The density function of the uniform distribution on the unit disc is given by

$$f(x, y) = \begin{cases} 1/\pi & \text{if } x^2 + y^2 \le 1, \\ 0 & \text{otherwise.} \end{cases}$$

Let g be the probability density function of the uniform distribution on the rectangle

$$[-1, 1] \times [-1, 1].$$

That is,

$$g(x, y) = \begin{cases} 1/4 & \text{if } -1 \le x, y \le 1, \\ 0 & \text{otherwise.} \end{cases}$$

The constant c is given by

$$c = \max_{-1 \le x, y \le 1} \frac{f(x, y)}{g(x, y)} = \frac{4}{\pi} \approx 1.273.$$

The acceptance probability of a trial sample (X, Y) from the density g is

$$\frac{1}{c}\frac{f(X,Y)}{g(X,Y)} = \begin{cases} 1 & \text{if } X^2 + Y^2 \leq 1, \\ 0 & \text{otherwise.} \end{cases}$$

Since $c \approx 1.273$, on average 1.273 samples from the uniform distribution on $[-1, 1] \times [-1, 1]$ are needed to generate one sample from the uniform distribution on the unit disc. Note that to generate a sample (X, Y) from the uniform distribution on $[-1, 1] \times [-1, 1]$, it suffices to generate X and Y independently from the uniform distribution on $[-1, 1]$.

Pseudocode:

($*$) generate two independent samples X and Y uniformly from $[-1, 1]$
       accept (X, Y) if $X^2 + Y^2 \leq 1$, otherwise reject (X, Y) and go to ($*$).

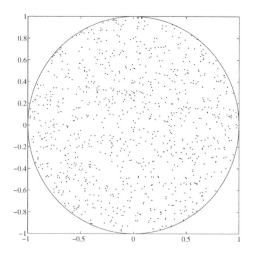

Figure 5.2: Uniform samples on a unit disc.

In the actual simulation, a total of 1255 trial samples are used to generate 1000 samples in the unit disc.   ∎

Example 5.4. Construct an acceptance-rejection scheme to generate samples from a Gamma distribution with density

$$f(x) = 2\sqrt{\frac{x}{\pi}}e^{-x}, \quad x \geq 0.$$

The alternative sampling distribution is assumed to be exponential. Which exponential distribution yields the most efficient sampling scheme?

SOLUTION: Suppose that the alternative sampling distribution is exponential with rate λ. That is, $g(x) = \lambda e^{-\lambda x}$ for $x \geq 0$. The smallest constant c that satisfies (5.1) is

$$c = \max_{x \geq 0} \frac{f(x)}{g(x)} = \max_{x \geq 0} \frac{2}{\sqrt{\pi}} \frac{\sqrt{x} e^{-x}}{\lambda e^{-\lambda x}}.$$

Note that the maximum is infinity if $\lambda \geq 1$. Therefore, we should only consider those $\lambda < 1$. In this case, the maximum is attained at $x^* = 0.5/(1 - \lambda)$ and

$$c = \sqrt{\frac{1}{2e\pi\lambda^2(1-\lambda)}}.$$

Since on average c samples from g are needed to generate a sample from f, the optimal λ^* should minimize c, or equivalently, maximize $\lambda^2(1 - \lambda)$. It follows that $\lambda^* = 2/3$, and the corresponding c is approximately 1.257.

Pseudocode:

($*$) generate a sample Y from the exponential distribution with rate $2/3$

generate a sample U from the uniform distribution on $[0, 1]$

accept Y if

$$U \leq \frac{1}{c} \frac{f(Y)}{g(Y)} = \sqrt{\frac{2eY}{3}} e^{-Y/3}$$

otherwise, discard Y and go to step ($*$).

The simulation results are presented in the histogram in Figure 5.3 with $\lambda = 2/3$. On average, 1.257 samples from the exponential distribution are needed to generate one sample of this Gamma distribution. The sample size is 10000, and in total 12488 trial samples are drawn from the exponential distribution. ■

(*) Look up

5.3 Sampling Multivariate Normal Distributions

Multivariate normal distributions are commonly used in financial engineering to model the joint distribution of multiple assets. Sampling from such distributions becomes less straightforward when the components are correlated; see the spread call option pricing problem considered in Example 4.3. In this section, we discuss a general scheme based on the Cholesky

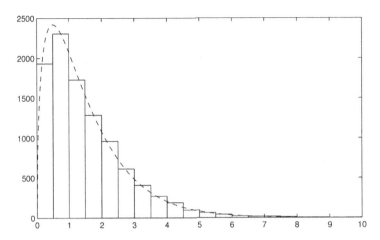

Figure 5.3: Samples of a Gamma distribution.

factorization. It is assumed that we are able to draw independent samples from the one-dimensional standard normal distribution; see Exercise 5.7.

Consider the d-dimensional multivariate normal distribution with mean μ and covariance matrix Σ. Without loss of generality, we assume that Σ is nonsingular (see Remark 5.1). Recall the following property. Let Z be a d-dimensional standard normal random vector. Then for any $d \times d$ matrix A,

$$X = \mu + AZ \tag{5.2}$$

is jointly normal with mean μ and covariance matrix AA'. Therefore, if one can

(a) sample from the d-dimensional standard normal distribution,

(b) find a matrix A such that $AA' = \Sigma$,

then (5.2) leads to an algorithm for generating samples from $N(\mu, \Sigma)$.

Note that (a) can be easily achieved. Indeed, if $Z_1, \ldots, Z_d$ are independent standard normal variables, then $Z = (Z_1, \ldots, Z_d)'$ is a d-dimensional standard normal random vector. The answer to (b) is not unique. There are many matrices A that satisfy $AA' = \Sigma$. A particularly convenient choice is given by the *Cholesky factorization* of Σ, which assumes that A is a lower triangular matrix:

$$A = \begin{bmatrix} A_{11} & 0 & 0 & \cdots & 0 \\ A_{21} & A_{22} & 0 & \cdots & 0 \\ A_{31} & A_{32} & A_{33} & \cdots & 0 \\ \vdots & \vdots & \vdots & \ddots & \vdots \\ A_{d1} & A_{d2} & A_{d3} & \cdots & A_{dd} \end{bmatrix}.$$

The advantage of choosing a lower triangular matrix is that one can derive explicit formulas for A_{ij}, and these formulas can be evaluated numerically by a simple recursion.

Omitting the details of derivation, we give the formula for a solution to this Cholesky factorization:

$$A_{11} = \sqrt{\Sigma_{11}}, \quad A_{j1} = \Sigma_{j1}/A_{11}, \quad j > 1,$$

$$A_{ii} = \sqrt{\Sigma_{ii} - \sum_{k=1}^{i-1} A_{ik}^2}, \quad A_{ji} = \left(\Sigma_{ji} - \sum_{k=1}^{i-1} A_{jk} A_{ik}\right)/A_{ii}, \quad j > i \geq 2.$$

Pseudocode for Cholesky factorization:

set $A = 0$ [$d \times d$ zero matrix]
set $v = 0$ [$d \times 1$ zero vector]
for $i = 1, 2, \ldots, d$
    for $j = i, \ldots, d$
        set $v_j = \Sigma_{ji}$
        for $k = 1, \ldots, i-1$
            set $v_j = v_j - A_{jk} A_{ik}$
        set $A_{ji} = v_j/\sqrt{v_i}$
return A.

Pseudocode for generating one sample from $N(\mu, \Sigma)$:

find a matrix A such that $AA' = \Sigma$ through Cholesky factorization
generate independent samples $Z_1, \ldots, Z_d$ from $N(0, 1)$
set $Z = (Z_1, \ldots, Z_d)'$
set $X = \mu + AZ$.

Example 5.5. Some option payoffs depend on multiple assets. Assume that under the risk-neutral probability measure the prices of these underlying assets are all geometric Brownian motions

$$S_t^{(i)} = S_0^{(i)} \exp\left\{ \left(r - \frac{1}{2}\sigma_i^2 \right) t + \sigma_i W_t^{(i)} \right\}, \quad i = 1, \dots, d,$$

where $W = (W^{(1)}, W^{(2)}, \cdots, W^{(d)})$ is a d-dimensional Brownian motion with covariance matrix $\Sigma = [\Sigma_{ij}]$ such that $\Sigma_{ii} = 1$ for all i. Consider an outperformance option with maturity T and payoff

$$\left(\max\left\{ c_1 S_T^{(1)}, \dots, c_d S_T^{(d)} \right\} - K \right)^+.$$

Estimate the price of this option.

SOLUTION: The price of the option is the expected value of the discounted payoff:

$$v = E\left[e^{-rT} \left(\max\left\{ c_1 S_T^{(1)}, \dots, c_d S_T^{(d)} \right\} - K \right)^+ \right].$$

In order to estimate v, we need to generate samples of $(S_T^{(1)}, \cdots, S_T^{(d)})$, or equivalently, those of $(W_T^{(1)}, \cdots, W_T^{(d)})$, which is a jointly normal random vector with mean 0 and covariance matrix $T\Sigma$.

Pseudocode:

> find a matrix A such that $AA' = \Sigma$ through Cholesky factorization
>
> for $i = 1, \dots, n$
>
> > generate independent samples $Z_1, \dots, Z_d$ from $N(0,1)$
> >
> > set $Z = (Z_1, \dots, Z_d)'$
> >
> > set $Y = AZ$
> >
> > for $k = 1, \dots, d$
> >
> > > set $S_k = S_0^{(k)} \exp\left\{ (r - \sigma_k^2/2)T + \sigma_k \sqrt{T} Y_k \right\}$
> >
> > set $H_i = e^{-rT} (\max\{ c_1 S_1, \dots, c_d S_d \} - K)^+$
>
> compute the estimate $\hat{v} = \frac{1}{n}(H_1 + \cdots + H_n)$
>
> compute the standard error S.E. $= \sqrt{\dfrac{1}{n(n-1)} \left(\sum_{i=1}^{n} H_i^2 - n\hat{v}^2 \right)}.$

The numerical results are reported in Table 5.1 for an outperformance option with $d = 4$ underlying assets. The parameters are given by

$$S_0^{(1)} = 45, \quad S_0^{(2)} = 50, \quad S_0^{(3)} = 45, \quad S_0^{(4)} = 55, \quad r = 0.02, \quad T = 0.5,$$

$$\sigma_1 = \sigma_2 = \sigma_3 = 0.1, \, \sigma_4 = 0.2, \, \Sigma = \begin{bmatrix} 1.0 & 0.3 & -0.2 & 0.4 \\ 0.3 & 1.0 & -0.3 & 0.1 \\ -0.2 & -0.3 & 1.0 & 0.5 \\ 0.4 & 0.1 & 0.5 & 1.0 \end{bmatrix}.$$

Table 5.1: Monte Carlo simulation for an outperformance option

	Sample size $n = 2500$			Sample size $n = 10000$		
Strike price K	50	55	60	50	55	60
M.C. Estimate	6.9439	3.5159	1.5406	7.0714	3.4197	1.4730
S.E.	0.1250	0.1009	0.0711	0.0624	0.0501	0.0346

The Cholesky factorization of Σ yields

$$A = \begin{bmatrix} 1.0000 & 0 & 0 & 0 \\ 0.3000 & 0.9539 & 0 & 0 \\ -0.2000 & -0.2516 & 0.9469 & 0 \\ 0.4000 & -0.0210 & 0.6069 & 0.6864 \end{bmatrix},$$

which satisfies $AA' = \Sigma$.   ∎

Remark 5.1. Let $X = (X_1, \ldots, X_d)'$ be a jointly normal random vector with mean μ and covariance matrix Σ. When Σ is singular, there exists a subset of the components of X whose covariance matrix is nonsingular, and such that every other component of X can be written as a linear combination of the components within this subset. Therefore, simulating X amounts to simulating this subset of components, which is a jointly normal random vector itself with a nonsingular covariance matrix.

Exercises

Pen-and-Paper Problems

5.1 Use the inverse transform method to generate samples from the following distributions. Write down the pseudocode.

(a) The uniform distribution on $[a, b]$.

(b) The exponential distribution with rate λ.

(c) The Weibull distribution with parameters (α, β) whose density is

$$f(x) = \begin{cases} \alpha\beta x^{\beta-1}e^{-\alpha x^\beta} & \text{if } x \geq 0, \\ 0 & \text{otherwise.} \end{cases}$$

(d) The Cauchy distribution whose density is

$$f(x) = \frac{1}{\pi}\frac{1}{(1+x^2)}, \quad x \in \mathbb{R}.$$

5.2 A random variable X is said to be *geometric with parameter p* if it takes values in $\{1, 2, \ldots\}$ and

$$\mathbb{P}(X = i) = (1-p)^{i-1}p, \quad i = 1, 2, \ldots.$$

Show that if Y is an exponential random variable with rate $\lambda = -\log(1-p)$, then (denote by $[x]$ the integer part of x)

$$X = 1 + [Y]$$

is geometric with parameter p. Use this observation to generate samples of X from the uniform distribution on $[0, 1]$. Write down the pseudocode.

5.3 Write down the pseudocode for generating samples of a discrete random variable with infinitely many possible values $x_1 < \cdots < x_n < \cdots$ and

$$\mathbb{P}(X = x_n) = p_n.$$

You may want to use the inverse transform method and the loop command "while" in MATLAB.

5.4 Consider a mixture probability distribution whose cumulative distribution function is

$$F(x) = \sum_{i=1}^{m} p_i F_i(x).$$

Here F_i is a cumulative distribution function itself for each i and $p_1, \ldots, p_m$ are some positive numbers such that

$$\sum_{i=1}^{m} p_i = 1.$$

Assume that we know how to generate samples from F_i for each i. Write down an algorithm for generating samples from the mixture F. *Hint:* Suppose that $Y_1, \ldots, Y_m$ are independent and the cumulative distribution function of Y_i is F_i for each i. Let I be an independent random variable such that $\mathbb{P}(I = i) = p_i$. What is the cumulative distribution function of Y_I?

5.5 Suppose that we wish to generate samples from a probability distribution with density

$$f(x) = \begin{cases} \frac{1}{2}x^2 e^{-x} & \text{if } x \geq 0, \\ 0 & \text{otherwise,} \end{cases}$$

by the acceptance-rejection method. The alternative density function is chosen to be

$$g(x) = \begin{cases} \lambda e^{-\lambda x} & \text{if } x \geq 0, \\ 0 & \text{otherwise,} \end{cases}$$

for some $\lambda \in (0, 1)$. Determine the best λ that minimizes the average number of trial samples needed to generate a sample from f.

5.6 Assume that X has density f. Design an acceptance-rejection scheme to draw samples of X conditional on $X > a$ for some given level a, with f as the alternative sampling density. Write down the pseudocode. On average, how many trial samples from f are needed to generate a desired sample? Do you think the assumption that X has a density is really necessary?

5.7 Let θ be uniform on $[0, 2\pi]$ and $R = \sqrt{2S}$ where S is an exponential random variable with rate one. Assume that θ and S are independent. Show that

$$X = R \cos \theta, \quad Y = R \sin \theta$$

are independent standard normal random variables. This is the *Box–Muller method* of generating standard normal random variables. *Hint:* Compute the joint density of X and Y using polar coordinates.

5.8 It happens quite often that one wishes to draw samples from a density function f that takes the form $f(x) = Ch(x)$, where h is a known nonnegative function and C is the *unknown* normalizing constant that satisfies

$$\int_{\mathbb{R}} f(x)\, dx = C \int_{\mathbb{R}} h(x)\, dx = 1.$$

Notationally, it is denoted by $f(x) \propto h(x)$. Let g be a density function from which one knows how to draw samples. Assume that for some positive constants k_1 and k_2,

$$k_1 \leq \frac{h(x)}{g(x)} \leq k_2$$

for every x. Show that

$$\max_x \frac{f(x)}{g(x)} \leq \frac{k_2}{k_1}.$$

Use this observation to design acceptance-rejection schemes to draw samples from the following distributions:

$$f(x) \propto e^{-x^2}(1 - e^{-\sqrt{1+x^2}}), \quad f(x) \propto (1 + e^{-x^2})(1 + x^2)^{-1}.$$

This method applies to higher dimensional distributions as well.

5.9 Argue that a Cholesky factorization for a 2×2 covariance matrix of the form

$$\Sigma = \begin{bmatrix} 1 & \rho \\ \rho & 1 \end{bmatrix}$$

is given by

$$A = \begin{bmatrix} 1 & 0 \\ \rho & \sqrt{1 - \rho^2} \end{bmatrix}.$$

That is, A is lower triangular and $AA' = \Sigma$. This is exactly the matrix C obtained in Example 4.3.

MATLAB® Problems

5.A Write a function that uses the inverse transform method to draw one sample of a discrete random variable X with

$$\mathbb{P}(X = i) = p_i, \quad i = 1, \ldots, m,$$

where p_i are all positive constants and $p_1 + \cdots + p_m = 1$. The function should include the parameters m and $p = (p_1, \ldots, p_m)$ as the input. Test your algorithm by generating 10000 samples for

$$m = 4, \quad p = (0.1, 0.2, 0.3, 0.4).$$

Compare the empirical frequencies with p.

5.B Assume that there are $(d + 1)$ underlying assets, whose prices under the risk-neutral probability measure are geometric Brownian motions

$$S_t^{(i)} = S_0^{(i)} \exp \left\{ \left(r - \frac{1}{2}\sigma_i^2 \right) t + \sigma_i W_t^{(i)} \right\}, \quad i = 1, \ldots, d + 1,$$

where $W = (W^{(1)}, \ldots, W^{(d+1)})$ is a $(d + 1)$-dimensional Brownian motion with covariance matrix $\Sigma = [\Sigma_{ij}]$ such that $\Sigma_{ii} = 1$ for every i. Write a function to estimate the price of an exchange option with maturity T and payoff

$$X = \left[\sum_{i=1}^{d} c_i S_T^{(i)} - S_T^{(d+1)} \right]^+.$$

The function should have input parameters $d, r, T, \sigma_1, \ldots, \sigma_{d+1}, c_1, \ldots, c_d, \Sigma,$ $S_0^{(1)}, \ldots, S_0^{(d+1)}$, and the sample size n. Report your estimate and standard error for the price of the option, given

$$d = 3, \quad r = 0.05, \quad T = 1, \quad c_1 = c_2 = c_3 = 1/3, \quad \sigma_1 = \sigma_2 = 0.2,$$

$$\sigma_3 = \sigma_4 = 0.3, \quad \Sigma = \begin{bmatrix} 1 & 0.1 & 0.2 & 0.2 \\ 0.1 & 1 & 0.3 & -0.3 \\ 0.2 & 0.3 & 1 & 0.5 \\ 0.2 & -0.3 & 0.5 & 1 \end{bmatrix},$$

$$S_0^{(1)} = S_0^{(2)} = S_0^{(3)} = 50, \quad S_0^{(4)} = 45, \quad n = 10000.$$

5.C Assume that under the risk-neutral probability measure the prices of two underlying assets are geometric Brownian motions

$$S_t^{(i)} = S_0^{(i)} \exp\left\{ \left(r - \frac{1}{2}\sigma_i^2 \right) t + \sigma_i W_t^{(i)} \right\}, \quad i = 1, 2,$$

where $W = (W^{(1)}, W^{(2)})$ is a two-dimensional Brownian motion with co-variance matrix

$$\Sigma = \begin{bmatrix} 1 & \rho \\ \rho & 1 \end{bmatrix}$$

for some $-1 < \rho < 1$. Consider a two-asset barrier option with maturity T and payoff

$$\left[S_T^{(1)} - K \right]^+ \cdot 1_{\left\{ \min(S_{t_1}^{(2)}, \ldots, S_{t_m}^{(2)}) \geq b \right\}}$$

where $0 < t_1 < \cdots < t_m = T$ are given dates. Write a function to estimate the price of this option. The function should have input parameters $S_0^{(1)}$, $S_0^{(2)}, r, \sigma_1, \sigma_2, \rho, K, b, T, m, (t_1, \ldots, t_m)$, and the sample size n. Report your estimate and standard error for

$$S_0^{(1)} = 50, \quad S_0^{(2)} = 40, \quad r = 0.03, \quad \sigma_1 = 0.2, \quad \sigma_2 = 0.4, \quad \rho = 0.2,$$

$$K = 50, \quad b = 38, \quad T = 1, \quad m = 50, \quad t_i = iT/m, \quad n = 10000.$$

5.D Suppose that under the risk-neutral probability measure, the stock price is a geometric Brownian motion with jumps:

$$S_t = S_0 \exp\left\{ \left(\bar{r} - \frac{1}{2}\sigma^2 \right) t + \sigma W_t + \sum_{i=1}^{N_t} Y_i \right\},$$

where W is a standard Brownian motion, $N = \{N_t : t \geq 0\}$ is a Poisson process with rate λ, and Y_i's are iid double exponential random variables with density

$$f(x) = p \cdot \eta_1 e^{-\eta_1 x} 1_{\{x \geq 0\}} + (1 - p) \cdot \eta_2 e^{\eta_2 x} 1_{\{x < 0\}}$$

for some $p \in (0,1)$ and positive constants η_1, η_2. Assume that W, N, and $\{Y_i\}$ are independent. The parameter $\bar{r}$ no longer equals the risk-free interest rate r. Instead, it should be chosen so that $E[e^{-\bar{r}T} S_T] = S_0$, or equivalently,

$$\bar{r} = r - \lambda \left(\frac{p\eta_1}{\eta_1 - 1} + \frac{(1-p)\eta_2}{\eta_2 + 1} - 1 \right);$$

see [17]. The price of an option with payoff X and maturity T still takes the form

$$v = E[e^{-rT} X],$$

where the expected value is taken under the risk-neutral probability measure. Write a function to price the call option with strike price K and maturity T. The function should have input parameters r, σ, λ, p, η_1, η_2, S_0, K, T, and the sample size n. Report your estimate and standard error for

$$r = 0.05, \ \sigma = 0.2, \ \lambda = 3, \ p = 0.4, \ \eta_1 = \eta_2 = 20,$$

$$S_0 = 50, \ K = 50, \ T = 0.5, \ n = 10000.$$

Chapter 6

Variance Reduction Techniques

The efficiency of a Monte Carlo estimate is often characterized by its variance. The smaller the variance, the more efficient the estimate. In this chapter, we will discuss some commonly used variance reduction techniques in Monte Carlo simulation, including antithetic sampling, the control variate method, and stratified sampling. These methods can also be combined to further improve efficiency. It should be noted that there is no "panacea" in this business of variance reduction. The method of choice is very much problem dependent.

6.1 Antithetic Sampling

In plain Monte Carlo simulation, samples are independent and identically distributed. The idea of antithetic sampling is to reduce the variance by introducing samples that are *negatively* correlated. To be more precise, consider the following two schemes for estimating $E[X]$. In both scenarios, the estimate is the sample average.

1. *Plain Monte Carlo:* $2n$ iid samples $X_1, \ldots, X_n, X_{n+1}, \ldots X_{2n}$. The estimate is

$$\frac{1}{2n} \sum_{i=1}^{2n} X_i.$$

2. *Antithetic Sampling:* $2n$ samples (or n pairs of samples)

$$\begin{matrix} X_1 & X_2 & \cdots & X_n \\ Y_1 & Y_2 & \cdots & Y_n \end{matrix}.$$

Pairs of samples (X_i, Y_i) are iid; Y_i has the same distribution as X_i; X_i and Y_i are *dependent*. The estimate is

$$\hat{v} = \frac{1}{n} \sum_{i=1}^{n} \frac{X_i + Y_i}{2}.$$

Let $\sigma^2 = \text{Var}[X_i]$. The plain Monte Carlo estimate is unbiased, and its variance is

$$\text{Var}\left(\frac{1}{2n} \sum_{i=1}^{2n} X_i \right) = \frac{1}{2n} \sigma^2.$$

Since Y_i has the same distribution as X_i, the antithetic sampling estimate $\hat{v}$ is again unbiased and

$$\text{Var}[\hat{v}] = \frac{1}{4n^2} \sum_{i=1}^{n} \text{Var}(X_i + Y_i).$$

Suppose that the correlation coefficient between X_i and Y_i is β. It follows that

$$\text{Var}[X_i + Y_i] = \text{Var}[X_i] + \text{Var}[Y_i] + 2\text{Cov}(X_i, Y_i) = 2\sigma^2 + 2\beta\sigma^2,$$

and

$$\text{Var}[\hat{v}] = \frac{1}{2n} \sigma^2 + \frac{\beta}{2n} \sigma^2.$$

Therefore, antithetic sampling achieves variance reduction when X_i and Y_i are *negatively* correlated, i.e., when $\beta < 0$. The improvement is characterized by the magnitude of β — the stronger the negative correlation, the more significant the variance reduction. Note that if the samples X_i and Y_i are made to be positively correlated, then antithetic sampling will actually increase the variance and make the estimate less accurate!

Remark 6.1. The standard deviation associated with the antithetic sampling estimate $\hat{v}$ is

$$\sqrt{\frac{1}{n} \text{Var}\left[\frac{X_i + Y_i}{2} \right]}.$$

Replacing the variance by sample variance, we obtain the standard error

$$\text{S.E.} = \sqrt{\frac{1}{n(n-1)} \left(\sum_{i=1}^{n} \left[\frac{X_i + Y_i}{2} \right]^2 - n\hat{v}^2 \right)}.$$

6.1.1   Generating Antithetic Samples

Now the question is: given a sample X_i, how should the antithetic sample Y_i be defined and generated? To ease exposition, we will omit the index i and denote the samples by X and Y, respectively.

In theory, it is always possible to construct a negatively correlated anti-thetic sample. Indeed, by Theorem 5.1 one can write $X = F^{-1}(U)$, where U is a random variable uniformly distributed on $[0, 1]$ and F^{-1} is the inverse of the cumulative distribution function of X. An antithetic sample of X can be defined as

$$Y = F^{-1}(1 - U).$$

Clearly Y has the same distribution as X since $1 - U$ is also uniformly dis-tributed on $[0, 1]$. Furthermore, X and Y are negatively correlated because X is an increasing function of U and Y is a decreasing function of U; see Exercise 6.4.

In practice, the inverse function F^{-1} is usually not available. However, the previous discussion does suggest that if $X = h(U)$ for some function h, then an antithetic sample can be defined as

$$Y = h(1 - U).$$

If h is monotone, then by the same token X and Y are negatively correlated. However, when h fails to be monotone, one has to exercise caution because the correlation between X and Y could be positive. We can also extend this discussion to the more general cases.

a. $X = h(Z)$ **where Z is $N(0, 1)$.** One can write $Z = \Phi^{-1}(U)$, and thus the antithetic sample is

$$Y = h(\Phi^{-1}(1 - U)) = h(-\Phi^{-1}(U)) = h(-Z).$$

b. $X = h(U_1, \ldots, U_k)$ **where $\{U_1, \ldots, U_k\}$ are iid uniform on $[0, 1]$.** The antithetic sample is

$$Y = h(1 - U_1, \ldots, 1 - U_k).$$

c. $X = h(Z_1, \ldots, Z_k)$ **where $\{Z_1, \ldots, Z_k\}$ are iid $N(0, 1)$.** The anti-thetic sample is

$$Y = h(-Z_1, \ldots, -Z_k).$$

Remark 6.2. Antithetic sampling does not exploit much knowledge of the underlying stochastic models, and thus its effectiveness is limited. It is of-ten used as part of a larger scheme to achieve greater variance reduction.

6.1.2 Examples of Antithetic Sampling

For all the examples in this section, we assume that the underlying stock price S is a geometric Brownian motion under the risk-neutral probability measure. That is,

$$S_t = S_0 \exp\left\{\left(r - \frac{1}{2}\sigma^2\right)t + \sigma W_t\right\},$$

where r is the risk-free interest rate.

Example 6.1. Use antithetic sampling to estimate the price of a call option with maturity T and strike price K. Compare with the plain Monte Carlo estimate.

SOLUTION: The call option payoff is an increasing function of S_T, and hence an increasing function of W_T. We expect that antithetic sampling will reduce the variance. The pseudocode for the plain Monte Carlo scheme is given in Example 4.1.

Pseudocode for antithetic sampling:

for $i = 1, 2, \ldots, n$
    generate a sample Z from $N(0, 1)$
    set $S_i = S_0 \exp\left\{\left(r - \frac{1}{2}\sigma^2\right)T + \sigma\sqrt{T}Z\right\}$
    set $X_i = e^{-rT}(S_i - K)^+$
    set $S_i = S_0 \exp\left\{\left(r - \frac{1}{2}\sigma^2\right)T - \sigma\sqrt{T}Z\right\}$
    set $Y_i = e^{-rT}(S_i - K)^+$
    compute the estimate $\hat{v} = \frac{1}{n}\sum_{i=1}^{n}\frac{X_i + Y_i}{2}$
    compute the standard error

$$\text{S.E.} = \sqrt{\frac{1}{n(n-1)}\left(\sum_{i=1}^{n}\left[\frac{X_i + Y_i}{2}\right]^2 - n\hat{v}^2\right)}.$$

The simulation results are given in Table 6.1 for call options with different strike prices. The parameters are

$$r = 0.05, \quad \sigma = 0.2, \quad T = 1, \quad S_0 = 50, \quad n = 10000.$$

The true values are obtained from the Black–Scholes formula. As expected, the antithetic sampling method does reduce the variance, but only to some extent.

Table 6.1: Call option: antithetic sampling versus plain Monte Carlo

Strike price	$K = 40$		$K = 50$		$K = 60$	
	antithetic	plain	antithetic	plain	antithetic	plain
True value	12.2944		5.2253		1.6237	
Estimate	12.2638	12.3376	5.2679	5.2142	1.6527	1.6251
S.E.	0.0231	0.0680	0.0372	0.0521	0.0287	0.0304

Note that for a give n, we use n pairs of samples (X_i, Y_i) for the antithetic sampling scheme and use $2n$ samples for the plain Monte Carlo scheme. ∎

Example 6.2. Consider a discretely monitored down-and-out barrier option with maturity T and payoff

$$(S_T - K)^+ \cdot 1_{\{\min(S_{t_1}, \cdots, S_{t_m}) \geq b\}}.$$

The monitoring dates $0 < t_1 < t_2 < \cdots < t_m = T$ are prefixed. Compare the antithetic sampling estimate with the plain Monte Carlo estimate.

SOLUTION: This is a path-dependent option and its payoff is monotonically increasing with respect to the stock price. Therefore, we expect that antithetic sampling will reduce the variance. Note that the sample path $(S_{t_1}, \ldots, S_{t_m})$ is generated sequentially by

$$S_{t_{i+1}} = S_{t_i} \exp\left\{ \left(r - \frac{1}{2}\sigma^2 \right)(t_{i+1} - t_i) + \sigma\sqrt{t_{i+1} - t_i}Z_{i+1} \right\},$$

where Z_i's are iid standard normal random variables. The antithetic sample path $(\bar{S}_{t_1}, \ldots, \bar{S}_{t_m})$ is given by

$$\bar{S}_{t_{i+1}} = \bar{S}_{t_i} \exp\left\{ \left(r - \frac{1}{2}\sigma^2 \right)(t_{i+1} - t_i) - \sigma\sqrt{t_{i+1} - t_i}Z_{i+1} \right\}$$

with the same initial price $\bar{S}_0 = S_0$. The pseudocode for the plain Monte Carlo scheme is very straightforward and thus omitted.

Pseudocode for antithetic sampling:

for $i = 1, 2, \ldots, n$

    for $j = 1, 2, \ldots, m$

        generate a sample Z from $N(0, 1)$

        set $S_j = S_{j-1} \exp\left\{ \left(r - \frac{1}{2}\sigma^2 \right) (t_j - t_{j-1}) + \sigma\sqrt{t_j - t_{j-1}} Z \right\}$

        set $\bar{S}_j = \bar{S}_{j-1} \exp\left\{ \left(r - \frac{1}{2}\sigma^2 \right) (t_j - t_{j-1}) - \sigma\sqrt{t_j - t_{j-1}} Z \right\}$

        set $X_i = e^{-rT} (S_m - K)^+ \cdot 1_{\{\min(S_1, \cdots, S_m) \geq b\}}$

        set $Y_i = e^{-rT} (\bar{S}_m - K)^+ \cdot 1_{\{\min(\bar{S}_1, \cdots, \bar{S}_m) \geq b\}}$

        compute the estimate $\hat{v} = \dfrac{1}{n} \sum\limits_{i=1}^{n} \dfrac{X_i + Y_i}{2}$

        compute the standard error

$$\text{S.E.} = \sqrt{\frac{1}{n(n-1)} \left(\sum_{i=1}^{n} \left[\frac{X_i + Y_i}{2} \right]^2 - n\hat{v}^2 \right)}.$$

The simulation results are reported in Table 6.2. The parameters are given by

$$r = 0.05, \quad \sigma = 0.2, \quad T = 1, \quad S_0 = 50, \quad b = 45,$$
$$m = 12, \quad t_i = iT/m, \quad n = 10000.$$

Table 6.2: Barrier option: antithetic sampling versus plain Monte Carlo

Strike price	$K = 40$		$K = 50$		$K = 60$	
	antithetic	plain	antithetic	plain	antithetic	plain
Estimate	9.8839	9.9773	4.7669	4.8529	1.5651	1.5672
S.E.	0.0423	0.0767	0.0399	0.0528	0.0283	0.0308

As expected, antithetic sampling reduces the variance to some extent. ∎

Example 6.3. Use antithetic sampling to estimate the price of a butterfly spread option with maturity T and payoff

$$(S_T - K_1)^+ + (S_T - K_3)^+ - 2(S_T - K_2)^+,$$

where $0 < K_1 < K_3$ and $K_2 = (K_1 + K_3)/2$. Compare with the plain Monte Carlo estimate.

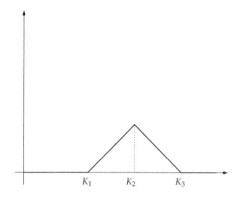

Butterfly spread option

SOLUTION: We omit the pseudocode since it is exactly the same as that of Example 6.1, except that the call option payoff should be replaced by the butterfly spread option payoff. The numerical comparison is presented in Table 6.3, where the parameters are

$$r = 0.05, \sigma = 0.2, S_0 = 50, K_1 = 45, K_2 = 50, K_3 = 55, n = 10000.$$

The true values are calculated from the Black–Scholes formula.

Table 6.3: Butterfly option: antithetic sampling versus plain Monte Carlo

Maturity	$T = 1$		$T = 0.5$		$T = 0.25$	
	antithetic	plain	antithetic	plain	antithetic	plain
True value	0.9192		1.3183		1.8156	
Estimate	0.9204	0.9180	1.3057	1.3373	1.8411	1.8311
S.E.	0.0123	0.0102	0.0152	0.0115	0.0166	0.0120

This example shows that when the payoff function is not monotone, the antithetic sampling method may increase the variance. ∎

6.2 Control Variates

Suppose that we are interested in estimating the expected value $E[X]$. The difference between the plain Monte Carlo scheme and the control variates method is as follows.

1. *Plain Monte Carlo:* n iid samples $X_1, \ldots, X_n$. The estimate is

$$\frac{1}{n} \sum_{i=1}^{n} X_i.$$

2. *Control Variates:* n samples $\{X_i\}$ and n control variate samples $\{Y_i\}$

$$\begin{array}{cccc} X_1 & X_2 & \cdots & X_n \\ Y_1 & Y_2 & \cdots & Y_n \end{array}.$$

Pairs of samples (X_i, Y_i) are iid; Y_i has *known* expected value $\bar{\mu}$. The estimate is

$$\hat{v} = \frac{1}{n} \sum_{i=1}^{n} X_i - b \left(\frac{1}{n} \sum_{i=1}^{n} Y_i - \bar{\mu} \right),$$

where b is a fixed constant.

Clearly, both estimates are unbiased. In general, the scheme of control variates is set up in such a way that sampling the pair (X_i, Y_i) requires little extra computational effort compared with sampling X_i alone.

Denote by σ_X^2 the variance of X_i, σ_Y^2 the variance of Y_i, and β the correlation coefficient between X_i and Y_i. The variance of the plain Monte Carlo estimate is

$$\frac{1}{n} \sigma_X^2.$$

As for the control variate estimate $\hat{v}$, write

$$\hat{v} = \frac{1}{n} \sum_{i=1}^{n} H_i, \quad H_i = X_i - b(Y_i - \bar{\mu}).$$

Since $H_1, \ldots, H_n$ are iid random variables, it follows that the variance of the control variate estimate equals

$$\frac{1}{n^2} \sum_{i=1}^{n} \text{Var}[H_i] = \frac{1}{n} \left[\sigma_X^2 - 2b\beta \sigma_X \sigma_Y + b^2 \sigma_Y^2 \right].$$

The size of variance reduction depends on the coefficient b. The *optimal* choice of b is the one that minimizes the variance of the control variate estimate:

$$b^* = \beta \frac{\sigma_X}{\sigma_Y} = \frac{\text{Cov}(X, Y)}{\text{Var}[Y]}. \tag{6.1}$$

If one uses b^*, the variance of the control variate estimate becomes

$$\frac{1}{n}(1 - \beta^2)\sigma_X^2.$$

In other words, it reduces the variance of the plain Monte Carlo estimate by a factor of β^2.

There are usually many possible ways to select the control variate Y. Ideally, it should have a strong correlation with X, be it positive or negative. For this reason, the control variate is often chosen to have a structure similar to X.

Remark 6.3. The standard deviation associated with the control variate estimate $\hat{v}$ is

$$\sqrt{\frac{1}{n}\text{Var}[H_i]}.$$

Replacing the variance by the sample variance, we obtain the standard error of $\hat{v}$:

$$\text{S.E.} = \sqrt{\frac{1}{n(n-1)}\left(\sum_{i=1}^{n} H_i^2 - n\hat{v}^2\right)}.$$

6.2.1 The Optimal Coefficient b^*

The optimal coefficient b^*, given by (6.1), is an unknown quantity in general. A common approach is to estimate b^* by the sample variance and covariance, using the same samples (X_i, Y_i). That is,

$$\hat{b}^* = \frac{\sum_{i=1}^{n}(X_i - \bar{X})(Y_i - \bar{Y})}{\sum_{i=1}^{n}(Y_i - \bar{Y})^2}, \quad \bar{X} = \frac{1}{n}\sum_{i=1}^{n} X_i, \quad \bar{Y} = \frac{1}{n}\sum_{i=1}^{n} Y_i. \tag{6.2}$$

Since the sample size n is large, the estimate $\hat{b}^*$ will be close to the true b^*. However, it should be pointed out that by using $\hat{b}^*$, we introduce bias. The reason is that in general

$$E\left[\hat{b}^*\left(\frac{1}{n}\sum_{i=1}^{n} Y_i - \mu\right)\right] \neq 0,$$

since $\hat{b}^*$ is no longer independent of $\{Y_i\}$. In practice, this bias is usually very small compared to the standard error of the estimate, and can be safely ignored.

If one wishes to eliminate this bias, one can estimate b^* from samples other than (X_i, Y_i) [such samples are sometimes called the *pilot samples*]. By using pilot samples, the estimate of b^* will be independent of $\{Y_i\}$, and hence the unbiasedness of the control variate estimate will be preserved. Typically, the size of pilot samples is much smaller than n so that it does not incur too much extra computational effort.

Yet another way to eliminate the bias is to forgo optimality and prefix b in a sensible way. For example, if X and Y are positively correlated and have *very* similar structures, one can let $b = 1$ or choose a b from prior experience; see Exercise 6.7. This naive approach can be quite effective sometimes.

6.2.2   Examples of Control Variates

For all the examples in this section, the underlying stock price S is assumed to be a geometric Brownian motion under the risk-neutral probability measure. That is,

$$S_t = S_0 \exp\left\{ \left(r - \frac{1}{2}\sigma^2 \right) t + \sigma W_t \right\},$$

where r is the risk-free interest rate.

Example 6.4. Underlying stock price as control variate. Use the method of control variates to estimate the price of a call option with maturity T and strike price K. Since

$$E[e^{-rT} S_T] = S_0,$$

the discounted stock price $e^{-rT} S_T$ can serve as a control variate.

SOLUTION: The control variate estimate is the sample average of iid copies of

$$H = e^{-rT}(S_T - K)^+ - b(e^{-rT} S_T - S_0).$$

We will either use $b = 1$ or the estimate $\hat{b}^*$ from formula (6.2).

Pseudocode for control variate method:

for $i = 1, 2, \ldots, n$

    generate Z from $N(0, 1)$

    set $S = S_0 \exp\left\{ \left(r - \frac{1}{2}\sigma^2 \right) T + \sigma\sqrt{T}Z \right\}$

    set $X_i = e^{-rT}(S - K)^+$

$$\text{set } Y_i = e^{-rT} S - S_0$$

set $b = 1$ or $b = \hat{b}^*$

for $i = 1, 2, \ldots, n$

    set $H_i = X_i - bY_i$

compute the estimate $\hat{v} = \dfrac{1}{n} \sum_{i=1}^{n} H_i$

compute the standard error S.E. $= \sqrt{\dfrac{1}{n(n-1)} \left(\sum_{i=1}^{n} H_i^2 - n\hat{v}^2 \right)}.$

The numerical results are shown in Table 6.4, where the parameters are given by

$$S_0 = 50, \quad r = 0.05, \quad \sigma = 0.2, \quad T = 1, \quad n = 10000.$$

Table 6.4: Call option: control variates versus plain Monte Carlo

Strike price	$K = 45$			$K = 55$		
	$b = 1$	$b = \hat{b}^*$	Plain MC	$b = 1$	$b = \hat{b}^*$	Plain MC
True value	8.3497			3.0200		
Estimate	8.3036	8.3104	8.3459	3.0572	3.0279	3.0001
S.E.	0.0262	0.0206	0.0860	0.0604	0.0308	0.0581

As we can see, the estimate with $b = \hat{b}^*$ is the most accurate one. Furthermore, the performance of the control variate estimates is better for the smaller strike price. The reason is that as K becomes smaller, the correlation between the payoff $(S_T - K)^+$ and the stock price S_T becomes stronger. ∎

Example 6.5. Analytically tractable derivatives as control variate. Estimate the price of a discretely monitored average price call option with maturity T and payoff $(\bar{S} - K)^+$. Here $\bar{S}$ is the arithmetic mean of stock prices:

$$\bar{S} = \frac{1}{m} \sum_{k=1}^{m} S_{t_k},$$

where $0 < t_1 < \cdots < t_m = T$ are given dates. Use the average price call option with geometric mean as the control variate.

SOLUTION: The price of an average price call option with geometric mean can be explicitly evaluated; see Example 2.5. Denote the price by p. Then the control variate estimate for the price of an average price call option with arithmetic mean is the sample average of iid copies of

$$e^{-rT} (\bar{S} - K)^+ - b \left[e^{-rT}(\bar{S}_G - K)^+ - p \right],$$

where

$$\bar{S}_G = \left(\prod_{k=1}^{m} S_{t_k} \right)^{1/m}.$$

Again, we let $b = 1$ or $b = \hat{b}^*$ from formula (6.2).

Pseudocode for control variate method:

for $i = 1, 2, \ldots, n$
  for $k = 1, 2, \ldots, m$
    generate Z from $N(0,1)$
    set $S_k = S_{k-1} \exp \left\{ \left(r - \frac{1}{2}\sigma^2 \right) (t_k - t_{k-1}) + \sigma\sqrt{t_k - t_{k-1}}Z \right\}$
  compute the arithmetic mean $\bar{S}$ of $S_1, \ldots, S_m$
  compute the geometric mean $\bar{S}_G$ of $S_1, \ldots, S_m$
  set $X_i = e^{-rT} (\bar{S} - K)^+$
  set $Y_i = e^{-rT} (\bar{S}_G - K)^+ - p$
set $b = 1$ or $b = \hat{b}^*$
for $i = 1, 2, \ldots, n$
  set $H_i = X_i - bY_i$
compute the estimate $\hat{v} = \frac{1}{n} \sum_{i=1}^{n} H_i$

compute the standard error S.E. $= \sqrt{\frac{1}{n(n-1)} \left(\sum_{i=1}^{n} H_i^2 - n\hat{v}^2 \right)}.$

We report the simulation results in Table 6.5. The parameters are given by

$$r = 0.05, \quad T = 1, \quad \sigma = 0.2, \quad m = 12, \quad t_i = iT/m, \quad S_0 = 50,$$

and the sample size is $n = 10000$.

Table 6.5: Asian option: control variates versus plain Monte Carlo

Strike price	$K = 45$			$K = 55$		
	$b = 1$	$b = \hat{b}^*$	plain	$b = 1$	$b = \hat{b}^*$	plain
Estimate	6.4590	6.4607	6.3755	1.1477	1.1456	1.1863
S.E.	0.0017	0.0013	0.0554	0.0018	0.0011	0.0288

The variance reduction achieved by the control variate method is signifi-cant even with the naive choice of $b = 1$. It reflects the strong correlation between the arithmetic mean $\bar{S}$ and the geometric mean $\bar{S}_G$.   ■

6.3 Stratified Sampling

Consider the problem of estimating $\mu = E[X]$. Suppose that there exists a random variable Y such that

1. Y takes finitely many possible values $\{y_1, \ldots, y_k\}$ and the probability mass function
$$p_i = \mathbb{P}(Y = y_i), \quad i = 1, \ldots, k$$
is known.

2. For each $i = 1, \ldots, k$, one is able to sample X *conditional* on $Y = y_i$.

The random variable Y partitions the sample space into k subsets according to its value. Each subset is called a *stratum*, and Y is said to be a *stratification variable*.

It follows from the tower property of conditional expectations (Theorem 1.12) that

$$\mu = E[X] = E[E[X|Y]] = \sum_{i=1}^{k} p_i E[X|Y = y_i]. \tag{6.3}$$

Since p_i is known, we only need to estimate $E[X|Y = y_i]$ for each i. This is made possible by the assumption that one can sample X conditional on Y. More precisely, a stratified sampling scheme divides the total sample size n into

$$n = n_1 + \cdots + n_k$$

and draws n_i samples $\{X_{ij} : j = 1, \ldots, n_i\}$ from the stratum determined by $Y = y_i$, that is, n_i samples of X conditional on $Y = y_i$. The stratified sampling estimate is defined to be

$$\hat{\mu} = \sum_{i=1}^{k} p_i \cdot \frac{1}{n_i} \sum_{j=1}^{n_i} X_{ij}. \tag{6.4}$$

Thanks to (6.3) and that $E[X_{ij}] = E[X|Y = y_i]$, the estimate $\hat{\mu}$ is clearly unbiased. Since X_{ij}'s are all independent, it follows that

$$\text{Var}[\hat{\mu}] = \sum_{i=1}^{k} p_i^2 \cdot \frac{1}{n_i^2} \sum_{j=1}^{n_i} \text{Var}[X_{ij}] = \sum_{i=1}^{k} p_i^2 \cdot \frac{1}{n_i} \sigma_i^2,$$

where

$$\sigma_i^2 = \text{Var}[X|Y = y_i].$$

Comments on stratified sampling.

- There are no fixed rules for choosing the stratification variable Y. In general, Y is selected in such a way that it is strongly correlated with X and the sampling of X conditional on $Y = y_i$ does not require too much additional computational effort.

- The sample size allocation $\{n_1, \ldots, n_k\}$ is assigned in advance. There are two commonly used approaches to select n_i, both of which lead to variance reduction. See Section 6.3.1.

- In general, the value of σ_i^2 is unknown. However, it can be estimated by the sample variance of $\{X_{ij} : j = 1, \ldots, n_i\}$:

$$s_i^2 = \frac{1}{n_i - 1} \sum_{j=1}^{n_i} \left(X_{ij} - \bar{X}_i\right)^2, \quad \bar{X}_i = \frac{1}{n_i} \sum_{j=1}^{n_i} X_{ij}.$$

Consequently, the standard error of $\hat{\mu}$ is just

$$\text{S.E.} = \sqrt{\sum_{i=1}^{k} p_i^2 \cdot \frac{1}{n_i} s_i^2}.$$

6.3.1   Allocating Samples in Each Stratum

The sample size in each stratum is often selected as a fixed fraction of the total sample size n. That is, $n_i = nq_i$, where

$$q_i > 0, \quad q_1 + \cdots + q_k = 1. \tag{6.5}$$

A very simple strategy is to let $q_i = p_i$, which is said to be the *proportional allocation*.

Another frequently used strategy aims to minimize the variance of $\hat{\mu}$. That is, one chooses the allocation $\{q_1, \ldots, q_k\}$ that minimizes

$$\text{Var}[\hat{\mu}] = \sum_{i=1}^{k} p_i^2 \cdot \frac{1}{n_i} \sigma_i^2 = \frac{1}{n} \sum_{i=1}^{k} p_i^2 \cdot \frac{1}{q_i} \sigma_i^2$$

under the constraints of (6.5). This optimization problem can be easily solved (see Remark 6.4), and the solution is the *optimal allocation*

$$q_i^* = \frac{p_i \sigma_i}{\sum_{m=1}^{k} p_m \sigma_m}. \tag{6.6}$$

In practice, since σ_i's are unknown, pilot samples are often used to obtain estimates for $\{\sigma_i\}$ and $\{q_i^*\}$.

Remark 6.4. The optimal allocation can be solved by the Cauchy–Schwartz inequality [14]: for any real numbers a_i and b_i

$$\sum_{i=1}^{k} a_i^2 \cdot \sum_{i=1}^{k} b_i^2 \geq \left(\sum_{i=1}^{k} a_i b_i \right)^2$$

with equality if and only if (assuming $b_i \neq 0$)

$$\frac{a_1}{b_1} = \cdots = \frac{a_k}{b_k}.$$

It follows that

$$\sum_{i=1}^{k} p_i^2 \frac{1}{q_i} \sigma_i^2 = \sum_{i=1}^{k} q_i \cdot \sum_{i=1}^{k} p_i^2 \frac{1}{q_i} \sigma_i^2 \geq \left(\sum_{i=1}^{k} p_i \sigma_i \right)^2$$

with equality if and only if

$$\frac{q_1}{p_1 \sigma_1} = \cdots = \frac{q_k}{p_k \sigma_k}.$$

Solving the equations in the last display along with the constraints in (6.5) leads to the optimal allocation (6.6).

6.3.2   Variance Decomposition

Compared with the plain Monte Carlo scheme, stratified sampling always achieves variance reduction with either the proportional allocation or the optimal allocation. We only need to verify this claim for the proportional allocation since by definition, the variance associated with the optimal allocation cannot exceed that of the proportional allocation.

Note that for the proportional allocation, $q_i = p_i$ and the variance of the corresponding estimate is

$$\sum_{i=1}^{k} p_i^2 \cdot \frac{1}{n_i} \sigma_i^2 = \frac{1}{n} \sum_{i=1}^{k} p_i \sigma_i^2.$$

Since the plain Monte Carlo estimate from n samples has variance $\mathrm{Var}[X]/n$, it suffices to show that

$$\mathrm{Var}[X] \geq \sum_{i=1}^{k} p_i \sigma_i^2. \tag{6.7}$$

To this end, let $\mu_i = E[X|Y = y_i]$ and observe that

$$E[X^2] = E[E[X^2|Y]] = \sum_{i=1}^{k} p_i E[X^2|Y = y_i] = \sum_{i=1}^{k} p_i (\sigma_i^2 + \mu_i^2),$$

$$E[X] = E[E[X|Y]] = \sum_{i=1}^{k} p_i E[X|Y = y_i] = \sum_{i=1}^{k} p_i \mu_i,$$

$$\mathrm{Var}[X] = E[X^2] - (E[X])^2 = \sum_{i=1}^{k} p_i \sigma_i^2 + \sum_{i=1}^{k} p_i \mu_i^2 - \left(\sum_{i=1}^{k} p_i \mu_i \right)^2.$$

The extra term

$$\sum_{i=1}^{k} p_i \mu_i^2 - \left(\sum_{i=1}^{k} p_i \mu_i \right)^2 \tag{6.8}$$

is nonnegative since it equals the variance of a discrete random variable that takes value μ_i with probability p_i. The desired inequality (6.7) follows readily.

Remark 6.5. The variance reduction achieved by the proportional allocation can also be understood from the variance decomposition formula

$$\mathrm{Var}[X] = E[\mathrm{Var}[X|Y]] + \mathrm{Var}(E[X|Y]),$$

which holds for arbitrary random variables X and Y (see Exercise 6.12). The variance from proportional allocation is indeed

$$\sum_{i=1}^{k} p_i \sigma_i^2 = E[\text{Var}[X|Y]],$$

while the extra term (6.8) equals $\text{Var}(E[X|Y])$. From this point of view, one can say that the proportional allocation eliminates the variance *between* strata, but not *within* strata.

6.3.3 Basic Strategies in Stratified Sampling

We should illustrate how one chooses the stratification variable Y and generates samples from the conditional distributions. The specific values that Y takes (that is, the y_i's) are not important. They are merely the indicators of strata. The essence of a stratification variable is a partition of the sample space.

The most basic algorithm in stratified sampling is the stratification of the uniform distribution on $[0,1]$. To be more concrete, suppose that one is interested in estimating $\mu = E[h(U)]$, where U is uniformly distributed on $[0,1]$ and h is an arbitrary function.

    a. **Stratification of Uniform Distribution on $[0,1]$.** A common strategy is to partition the unit interval into k strata of equal length:

$$I_1 = \left[0, \frac{1}{k}\right), \quad \cdots, \quad I_k = \left[\frac{k-1}{k}, 1\right].$$

Define Y so that $Y = i$ if and only if $U \in I_i$. It follows that for each i

$$p_i = \mathbb{P}(Y = i) = \mathbb{P}(U \in I_i) = \frac{1}{k}.$$

    b. **Stratified Sampling Estimate.** Let n_i be the sample size for the stratum $\{Y = i\} = \{U \in I_i\}$. Denote by $\{U_{ij} : j = 1, \ldots, n_i\}$ the iid samples from the conditional distribution of U given $U \in I_i$. The estimate is

$$\hat{\mu} = \sum_{i=1}^{k} \frac{1}{k} \cdot \frac{1}{n_i} \sum_{j=1}^{n_i} h(U_{ij}).$$

c. **Sample from Conditional Distributions.** It is trivial that conditional on $U \in I_i$, U is uniformly distributed on the interval I_i. Sampling from this uniform conditional distribution is straightforward. Indeed, let V_{ij} be iid samples from the uniform distribution on $[0, 1]$. Then

$$U_{ij} = \frac{i-1}{k} + \frac{1}{k}V_{ij}$$

are iid samples from the uniform distribution on I_i.

The stratified sampling scheme for estimating μ is now ready. Below is the pseudocode.

Pseudocode for stratified sampling:

specify the sample size n_i in stratum I_i for each i
for $i = 1, 2, \ldots, k$

generate iid samples $\{V_{ij} : j = 1, \ldots, n_i\}$ uniformly from $[0, 1]$
set $U_{ij} = (i-1)/k + V_{ij}/k$ for $j = 1, \ldots, n_i$
compute the sample mean and standard deviation in stratum I_i

$$\hat{\mu}_i = \frac{1}{n_i} \sum_{j=1}^{n_i} h(U_{ij}), \quad s_i = \sqrt{\frac{1}{n_i - 1} \sum_{j=1}^{n_i} [h(U_{ij}) - \hat{\mu}_i]^2}$$

compute the stratified sampling estimate $\hat{\mu} = \frac{1}{k} \sum_{i=1}^{k} \hat{\mu}_i$

compute the standard error S.E. $= \frac{1}{k} \sqrt{\sum_{i=1}^{k} \frac{1}{n_i} s_i^2}$.

Now consider the more general problem of estimating $E[h(X)]$. Denote by F the cumulative distribution function of X. By the inverse transform method, one can write $E[h(X)] = E[\bar{h}(U)]$, where $\bar{h} = h \circ F^{-1}$ and U is a random variable uniformly distributed on $[0, 1]$. Thus we have reverted to the previous setting. For this strategy to work, it is necessary that F^{-1} can be evaluated in a very efficient manner. For many financial engineering problems, this is not difficult because X can be chosen as a standard normal random variable and the inverse of its cumulative distribution function can be evaluated by the MATLAB® built-in function "norminv".

Example 6.6. Assume that under the risk-neutral probability measure the stock price is a geometric Brownian motion

$$S_t = S_0 \exp\left\{\left(r - \frac{1}{2}\sigma^2\right)t + \sigma W_t\right\}.$$

Design a stratified sampling scheme to estimate the price of a call option with maturity T and strike price K.

SOLUTION: The payoff is a function of S_T and thus a function of W_T. Write

$$W_T = \sqrt{T}\Phi^{-1}(U)$$

for some random variable U that is uniformly distributed on $[0, 1]$. Then the price of the call option is $v = E[h(U)]$, where

$$h(u) = \left(S_0 \exp\left\{-\frac{1}{2}\sigma^2 T + \sigma\sqrt{T}\Phi^{-1}(u)\right\} - e^{-rT}K\right)^+.$$

The pseudocode for stratified sampling is exactly the same as the one given in the preceding discussion. In the numerical simulation, we use proportional allocation and let

$$S_0 = 50, \quad r = 0.05, \quad \sigma = 0.2, \quad T = 1, \quad n = 10000.$$

The results are reported in Table 6.6.

Table 6.6: Stratified sampling for call option

Strike price	$K = 40$		$K = 50$		$K = 60$	
# of strata	$k = 25$	$k = 100$	$k = 25$	$k = 100$	$k = 25$	$k = 100$
True value	12.2944		5.2253		1.6237	
Estimate	12.2972	12.2979	5.2340	5.2358	1.6334	1.6279
S.E.	0.0127	0.0061	0.0124	0.0072	0.0136	0.0057

Compared with the results from the plain Monte Carlo scheme in Example 4.1, stratified sampling reduces the variance significantly.   ∎

Example 6.7. Use stratified sampling to estimate the price of a spread call option with maturity T and payoff

$$(X_T - Y_T - K)^+,$$

where X and Y are the prices of two underlying assets. Assume that under the risk-neutral probability measure,

$$X_t = X_0 \exp\left\{ \left(r - \frac{1}{2}\sigma_1^2 \right) t + \sigma_1 W_t \right\},$$

$$Y_t = Y_0 \exp\left\{ \left(r - \frac{1}{2}\sigma_2^2 \right) t + \sigma_2 B_t \right\},$$

where (W, B) is a two-dimensional Brownian motion with covariance matrix

$$\Sigma = \begin{bmatrix} 1 & \rho \\ \rho & 1 \end{bmatrix}.$$

SOLUTION: The payoff is a function of (X_T, Y_T), which is in turn a function of (θ, η), where

$$\theta = \frac{W_T}{\sqrt{T}}, \quad \eta = \frac{B_T}{\sqrt{T}}.$$

By assumption, (θ, η) is a jointly normal random vector with mean 0 and covariance matrix Σ. There are many ways to define a stratification variable. We will discuss two of them: (1) stratification of the random variable θ alone; (2) simultaneous stratification of the random vector (θ, η). In both cases we use proportional allocation.

Stratification of θ alone: In this approach, we first perform stratified sampling on θ. Since θ is a standard normal random variable, this can be done as in Example 6.6. The second step is to sample η conditional on the value of θ. This is not difficult because given $\theta = x$, η is normally distributed with mean ρx and variance $1 - \rho^2$; see Appendix A.

Pseudocode for stratifying θ alone:

specify the sample size n_i in stratum I_i for each i
for $i = 1, 2, \ldots, k$
    for $j = 1, 2, \ldots, n_i$
        generate a sample V_{ij} from the uniform distribution on $[0, 1]$
        set $U_{ij} = (i - 1)/k + V_{ij}/k$
        set $\theta_{ij} = \Phi^{-1}(U_{ij})$
        generate a sample η_{ij} from $N(\rho\theta_{ij}, 1 - \rho^2)$
        set H_{ij} as the discounted option payoff given (θ_{ij}, η_{ij})

compute the sample mean and standard deviation in stratum I_i

$$\hat{\mu}_i = \frac{1}{n_i} \sum_{j=1}^{n_i} H_{ij}, \quad s_i = \sqrt{\frac{1}{n_i - 1} \sum_{j=1}^{n_i} (H_{ij} - \hat{\mu}_i)^2}$$

compute the stratified sampling estimate $\hat{\mu} = \frac{1}{k} \sum_{i=1}^{k} \hat{\mu}_i$

compute the standard error S.E. $= \frac{1}{k} \sqrt{\sum_{i=1}^{k} \frac{1}{n_i} s_i^2}.$

The simulation results are reported in Table 6.7 with sample size $n = 10000$ for

$$X_0 = 50, \quad Y_0 = 45, \quad r = 0.05, \quad \sigma_1 = 0.2, \quad \sigma_2 = 0.3, \quad \rho = 0.5, \quad T = 1.$$

Table 6.7: Stratified sampling for spread call option

Strike price	$K = 0$		$K = 5$		$K = 10$	
# of strata	$k = 25$	$k = 100$	$k = 25$	$k = 100$	$k = 25$	$k = 100$
Estimate	7.9277	7.8958	4.8622	4.9584	2.7980	2.7988
S.E.	0.0782	0.0781	0.0630	0.0623	0.0484	0.0476

Compared with Example 4.3, the improvement of stratified sampling over plain Monte Carlo is almost negligible. The reason is that the stratification will only eliminate the variance *between strata*. In this case, since θ and η are positively correlated, so are X_T and Y_T. Therefore, the difference $X_T - Y_T$, or more precisely, the average of the difference $X_T - Y_T$ in each stratum, has little variation across the strata. Thus the variance reduction through stratification is inconsequential. One can also stratify η instead θ. The simulation result will only be slightly better.

Simultaneous stratification of (θ, η): Now let us consider another approach where we stratify θ and η simultaneously. To do this, recall from Example 4.3 that

$$CC' = \Sigma, \quad C = \begin{bmatrix} 1 & 0 \\ \rho & \sqrt{1 - \rho^2} \end{bmatrix}$$

is a Cholesky factorization of Σ. Suppose that U and V are independent and uniformly distributed on $[0, 1]$. Then

$$Z = \begin{bmatrix} \Phi^{-1}(U) \\ \Phi^{-1}(V) \end{bmatrix}$$

is a two-dimensional standard normal random vector and CZ is jointly normal with mean 0 and covariance matrix $CC' = \Sigma$. The scheme stratifies both U and V simultaneously by partitioning the interval $[0, 1]$ into k subintervals of equal length for U and for V. That is, there will be $k \times k = k^2$ strata $\{I_{ij}\}$ with

$$I_{ij} = \left[\frac{i-1}{k}, \frac{i}{k}\right) \times \left[\frac{j-1}{k}, \frac{j}{k}\right), \quad i, j = 1, \ldots, k.$$

To generate a sample from stratum I_{ij}, we generate u and v uniformly from $[0, 1]$ and let

$$U = (i-1)/k + u/k, \quad V = (j-1)/k + v/k.$$

Then we let

$$\begin{bmatrix} \theta \\ \eta \end{bmatrix} = C \begin{bmatrix} \Phi^{-1}(U) \\ \Phi^{-1}(V) \end{bmatrix}.$$

The pseudocode is straightforward and thus omitted. In order to match the number of strata in the previous simulation, we let $k = 5$ and $k = 10$ so that the total number of strata is $5^2 = 25$ and $10^2 = 100$, respectively. The simulation results are given in Table 6.8. We can see that this stratification strategy outperforms the previous one and the plain Monte Carlo scheme.

Table 6.8: Stratified sampling for spread call option

Strike price	$K = 0$		$K = 5$		$K = 10$	
# of strata k^2	$k = 5$	$k = 10$	$k = 5$	$k = 10$	$k = 5$	$k = 10$
Estimate	7.8979	7.8805	4.9766	4.9542	2.7966	2.8041
S.E.	0.0241	0.0139	0.0228	0.0137	0.0208	0.0129

It is worth noting that determining a good stratification variable is not always straightforward. Only in very specialized settings can one explicitly solve for the optimal stratification strategy. For more details, see [12]. ∎

Exercises

Pen-and-Paper Problems

6.1 Determine the antithetic variable and express it in terms of the original random variable X.

(a) X is normally distributed with mean μ and variance σ^2.

(b) X is lognormally distributed with parameters (μ, σ^2).

(c) X is uniformly distributed on the interval $[a, b]$.

(d) X is exponentially distributed with rate λ.

(e) X has a symmetric density function. That is, the density function f satisfies $f(x) = f(-x)$ for all x.

6.2 Suppose that X is a Bernoulli random variable with parameter p. Show that if $2p > 1$, then the antithetic variable of X is

$$Y = 1 - XZ,$$

where Z is a Bernoulli random variable with parameter $(1 - p)/p$, independent of X. What is the antithetic variable when $2p < 1$ or $2p = 1$?

6.3 Suppose that X and Y have the same distribution (X and Y could be dependent). Let $0 \leq \alpha \leq 1$ and define

$$Z = \alpha X + (1 - \alpha)Y.$$

Show that the variance of Z is minimized when $\alpha = 1/2$. This explains why in antithetic sampling $(X + Y)/2$ is used as the estimate instead of the more general $\alpha X + (1 - \alpha)Y$.

6.4 Let X be an arbitrary random variable. Let f be a monotonically increasing function and g a monotonically decreasing function. Show that $f(X)$ and $g(X)$ are negatively correlated. That is,

$$\text{Cov}[f(X), g(X)] \leq 0.$$

Hint: Let Y be an independent random variable that has the same distribution as X. Argue that $[f(X) - f(Y)] \cdot [g(X) - g(Y)] \leq 0$ and then take expected value on both sides.

6.5 Use antithetic sampling to estimate $P(Z \geq b)$, where Z is a standard normal random variable and b is a positive constant.

(a) Write down the antithetic sampling estimate.

(b) Compute the magnitude of variance reduction.

6.6 Construct an antithetic sampling algorithm to estimate $E[\exp\{U\}]$, where U is uniformly distributed on $[0,1]$. Compute the magnitude of variance reduction.

6.7 Show that the control variate method with $b = 1$ reduces the variance if and only if

$$2\beta\sigma_X > \sigma_Y.$$

6.8 When a random variable X has a known expected value μ, it can be used as the control variate in the estimation of $\mathbb{P}(X \leq a)$ for some given constant a. The estimate is the sample average of iid copies of

$$1_{\{X \leq a\}} - b(X - \mu),$$

for some constant b. Determine the optimal b and the magnitude of variance reduction when

(a) X is uniformly distributed on $[0,1]$ and $a \in (0,1)$;

(b) X is a standard normal random variable and $a \in \mathbb{R}$.

6.9 Suppose that one would like to estimate the price of a discretely monitored lookback put option with maturity T and payoff

$$X = \left(K - \min_{i=1,\dots,m} S_{t_i}\right)^+,$$

where $0 < t_1 < \cdots < t_m = T$ are given dates. The stock price is assumed to be a geometric Brownian motion under the risk-neutral probability measure. Which one of the following will you choose as the control variate? Explain your reasoning.

(a) $Y = e^{-rT}S_T$ with $\bar{\mu} = E[Y] = S_0$.

(b) $Y = e^{-rT}(K - S_T)^+$ with $\bar{\mu} = E[Y] = \text{BLS\_Put}$.

(c) $Y = e^{-rT}\left(K - \min_{0 \leq t \leq T} S_t\right)^+$, where $\bar{\mu} = E[Y]$ can be explicitly calculated; see Appendix C.

6.10 Let U be a random variable uniformly distributed on $[0,1]$. Consider the problem of estimating $E[U^2]$ by stratified sampling. Stratify U by partitioning the interval $[0,1]$ into $k = 2$ strata of equal length.

(a) Write down the estimate with the proportional allocation.

(b) Find the optimal allocation and write down the corresponding estimate.

Evaluate the magnitude of variance reduction from these two estimates.

6.11 Let X be exponentially distributed with rate λ. Design a stratified sampling scheme to estimate $\mu = E[h(X)]$. Write down the pseudocode.

6.12 Let X and Y be two arbitrary random variables. Define the *conditional variance* of X given Y by

$$\text{Var}[X|Y] = E[X^2|Y] - (E[X|Y])^2.$$

Show that $\text{Var}[X|Y] \geq 0$ and the variance decomposition formula

$$\text{Var}[X] = E[\text{Var}[X|Y]] + \text{Var}[E[X|Y]]$$

holds.

6.13 The variance decomposition formula in Exercise 6.12 suggests the following variance reduction technique. Consider the problem of estimating $\mu = E[h(X)]$. Let Y be a random variable for which

$$f(Y) = E[h(X)|Y]$$

is explicitly known. Let $\{Y_1, \ldots, Y_n\}$ be iid copies of Y, and define the estimate to be

$$\hat{\mu} = \frac{1}{n} \sum_{i=1}^{n} f(Y_i).$$

Show that $\hat{\mu}$ is unbiased and its variance is always no greater than the variance of the plain Monte Carlo estimate with the same sample size. This variance reduction technique is called the **method of conditioning**. Discuss the difference between the method of conditioning and stratified sampling.

6.14 Let (X, Y) be a jointly normal random vector with mean 0 and covariance matrix

$$\begin{bmatrix} 1 & \rho \\ \rho & 1 \end{bmatrix}.$$

Write down explicitly the estimate by the method of conditioning (conditional on Y) for

(a) $\mu = \mathbb{P}(XY \geq a)$ for some $a \in \mathbb{R}$;
(b) $\mu = \mathbb{P}(\min\{X, Y\} \geq a)$ for some $a \in \mathbb{R}$;
(c) $\mu = E[(e^X - e^Y)^+]$.

MATLAB® Problems

In Exercises 6.A – 6.D, assume that the underlying stock price is a geometric Brownian motion under the risk-neutral probability measure:

$$S_t = S_0 \exp\left\{ \left(r - \frac{1}{2}\sigma^2 \right) t + \sigma W_t \right\},$$

where W is a standard Brownian motion and r is the risk-free interest rate.

6.A Design simulation schemes to estimate the price of the butterfly option considered in Example 6.3.

(a) Control variate method: use the underlying stock price as the control variate and let $b = \hat{b}^*$, the sample estimate of b^*.

(b) Stratified sampling: stratify W_T into $k = 100$ strata and use the proportional allocation.

Report your simulation results with sample size $2n$. The parameters are the same as those given in Example 6.3.

6.B Consider a straddle with strike price K and maturity T. The option payoff is

$$X = (S_T - K)^+ + (K - S_T)^+.$$

Write a function to compare the following schemes:

(a) Plain Monte Carlo with sample size $2n$.

(b) Antithetic sampling with n pairs of samples.

(c) Control variate method with sample size $2n$. Use the underlying stock price as the control variate. Let $b = \hat{b}^*$.

(d) Stratified sampling with sample size $2n$. Stratify W_T into k strata and use the proportional allocation.

The function should have input parameters S_0, r, σ, K, T, k, and n. Report your estimates and standard errors for

$$S_0 = 50, \ r = 0.02, \ \sigma = 0.2, \ K = 50, \ T = 1, \ k = 100, \ n = 10000.$$

Why is the performance from the antithetic sampling and the control variate method poor?

6.C Consider a discretely monitored lookback call option with fixed strike price K and maturity T. The option payoff is

$$X = \left(\max_{i=1,\dots,m} S_{t_i} - K \right)^+,$$

where $0 < t_1 < \cdots < t_m = T$ are given dates. Write a function to compare the performance of the following simulation schemes.

(a) Plain Monte Carlo with sample size $2n$.

(b) Antithetic sampling with n pairs of samples.

(c) Control variate method with sample size $2n$. Use $b = \hat{b}^*$, the sample estimate of b^*. The control variate is

i. the underlying stock price;

ii. the call option with strike price K and maturity T;

iii. the discretely monitored average price call option with geometric mean and maturity T, whose payoff is

$$Y = (\bar{S}_G - K)^+, \quad \bar{S}_G = \left(\prod_{i=1}^{m} S_{t_i}\right)^{1/m};$$

iv. the continuous time lookback call option with strike price K and maturity T, whose payoff is

$$Y = \left(\max_{0 \le t \le T} S_t - K\right)^+.$$

Hint: You will need to sample from $(S_{t_1}, \ldots, S_{t_m}, \max_{0 \le t \le T} S_t)$. Use Exercise 4.7 (d).

The function should have input parameters $S_0, r, \sigma, K, T, m, (t_1, \ldots, t_m)$, and n. Report your estimates and standard errors for

$$S_0 = 50, \ r = 0.05, \ \sigma = 0.2, \ K = 55, \ T = 1,$$

$$m = 50, \ t_i = iT/m, \ n = 10000.$$

6.D Antithetic sampling and stratified sampling can be combined. For example, suppose that one is interested in estimating $\mu = E[h(Z)]$, where Z is a standard normal random variable. Observe that antithetic sampling is equivalent to estimating $\mu = E[\bar{h}(Z)]$ by the plain Monte Carlo method, where

$$\bar{h}(Z) = \frac{1}{2}[h(Z) + h(-Z)].$$

Now one can use stratified sampling to further improve the performance. Write a function to estimate the price of call options, combining antithetic sampling and stratified sampling. The function should have input parameters r, σ, T, K, S_0, the number of strata k, and the sample size n. Report your simulation results for

$$r = 0.05, \ \sigma = 0.2, \ T = 1, \ K = 50, \ S_0 = 50, \ n = 10000,$$

and $k = 25, 100$, respectively. Use the proportional allocation.

In Exercises 6.E – 6.G, assume that the prices of the two underlying stocks are both geometric Brownian motions under the risk-neutral probability measure:

$$X_t = X_0 \exp\left\{\left(r - \frac{1}{2}\sigma_1^2\right) T + \sigma_1 W_t\right\},$$

$$Y_t = Y_0 \exp\left\{\left(r - \frac{1}{2}\sigma_2^2\right) T + \sigma_2 B_t\right\}.$$

Again, r is the risk-free interest rate, whereas (W, B) is a two-dimensional Brownian motion with covariance matrix

$$\begin{bmatrix} 1 & \rho \\ \rho & 1 \end{bmatrix}.$$

6.E Suppose that we wish to estimate the price of a spread call option with maturity T and payoff

$$H = (X_T - Y_T - K)^+.$$

Let $Z_1 = W_T/\sqrt{T}$ and $Z_2 = B_T/\sqrt{T}$.

(a) What is the joint distribution of (Z_1, Z_2)?

(b) Given $Z_2 = z$, what is the distribution of Z_1?

(c) What is the conditional expected value $E[H|Z_2 = z]$?

(d) Write a function to estimate the option price by the method of conditioning; see Exercise 6.13. The function should have input parameters r, σ_1, σ_2, ρ, X_0, Y_0, T, K, and the sample size n. Report your estimate and standard error for

$$X_0 = 50, \ Y_0 = 45, \ r = 0.05, \ \sigma_1 = 0.2, \ \sigma_2 = 0.3, \ \rho = 0.5, \ T = 1,$$

with $n = 10000$ and $K = 0, 5, 10$, respectively. Compare your results with those of Example 6.7.

6.F Suppose that we wish to estimate the price of a basket call option with maturity T and payoff

$$H = (c_1 X_T + c_2 Y_T - K)^+.$$

Let $Z_1 = W_T/\sqrt{T}$ and $Z_2 = B_T/\sqrt{T}$. Write a function to compare the following simulation schemes.

(a) Method of conditioning (conditional on Z_2). Be careful with the sign of $c_2 Y_T - K$ when calculating $E[H|Z_2 = z]$.

(b) Stratified sampling. Always use the proportional allocation.

    i. Stratify Z_1 alone into k^2 strata.

    ii. Stratify Z_2 alone into k^2 strata.

    iii. Stratify Z_1 and Z_2 simultaneously into $k \times k = k^2$ strata.

The function should have input parameters r, σ_1, σ_2, ρ, X_0, Y_0, T, K, c_1, c_2, k, and the sample size n. Report your estimates and standard errors for

$$r = 0.1, \ \sigma_1 = 0.2, \ \sigma_2 = 0.3, \ X_0 = 50, \ Y_0 = 50, \ T = 1, \ K = 55,$$

$$\rho = 0.7, \ c_1 = 0.5, \ c_2 = 0.5$$

with $k = 10$ and sample size $n = 10000$.

6.G Consider the problem of estimating the price of a two-asset barrier option with maturity T and payoff

$$(X_T - K)^+ \cdot 1_{\{\min(Y_{t_1}, \dots, Y_{t_m}) \geq h\}},$$

where $0 < t_1 < \cdots < t_m = T$ are prefixed dates.

(a) Write a function to estimate the option price by the control variate method. Let $(X_T - K)^+$ be the control variate and use $b = \hat{b}^*$.

(b) Write a function to estimate the option price by the method of conditioning, conditional on $(Y_{t_1}, \dots, Y_{t_m})$ or equivalently $(B_{t_1}, \dots, B_{t_m})$. *Hint:* Use Exercise 2.16 to argue that one can write $W = \rho B + \sqrt{1 - \rho^2} Q$, where Q and B are independent standard Brownian motions. Use this to compute

$$E\left[(X_T - K)^+ \cdot 1_{\{\min(Y_{t_1}, \dots, Y_{t_m}) \geq h\}} \,\middle|\, Y_{t_1}, \dots, Y_{t_m} \right].$$

Compare your estimates and standard errors with those of Exercise 5.C for

$$X_0 = 50, \ Y_0 = 40, \ r = 0.03, \ \sigma_1 = 0.2, \ \sigma_2 = 0.4, \ \rho = 0.2, \ K = 50,$$

$$h = 38, \ T = 1, \ m = 50, \ t_i = iT/m, \ n = 10000.$$

Chapter 7

Importance Sampling
— Drawing samples from an alternate distribution (handwritten)

Importance sampling is a frequently used variance reduction technique in Monte Carlo simulation. It is particularly powerful for estimating small probabilities or expected values that are largely determined by events of small probabilities. In a nutshell, importance sampling draws samples from an alternative sampling distribution and compensates by multiplying the outcome with appropriate likelihood ratios.

The systematic construction of efficient importance sampling schemes requires advanced mathematical knowledge that is beyond the scope of this book. For this reason, we will limit our discussion to the basic ideas and rudimentary strategies of importance sampling, as well as the cross-entropy method, which is a simulation-based technique for selecting alternative sampling distributions. We will also address applications of importance sampling to option pricing and risk analysis.

7.1   Basic Ideas of Importance Sampling

Importance sampling is often referred to as a *change of measure* technique. To be more concrete, consider the problem of estimating the expected value

$$\mu = E[h(X)].$$

Suppose that X has a density $f(x)$. The plain Monte Carlo scheme will simulate iid samples $\{X_i\}$ from the density f and take the sample average of $\{h(X_i)\}$ as the estimate.

The basic idea of importance sampling comes from the observation that for an *arbitrary* density function $g(x)$ [see Remark 7.1], one can write

HW 7
1a)
$$\mu = \int_{\mathbb{R}} h(x) f(x) \, dx = \int_{\mathbb{R}} h(x) \frac{f(x)}{g(x)} g(x) \, dx = E\left[h(Y) \frac{f(Y)}{g(Y)}\right],$$

where Y is a random variable with density g. Therefore, one can draw iid samples $\{Y_i\}$ from the alternative density g and use the sample average of

$$h(Y_i) \frac{f(Y_i)}{g(Y_i)}$$

to estimate μ. Importance sampling is different from plain Monte Carlo in that samples are now generated from a different probability distribution (hence the name "change of measure") and in order to preserve unbiasedness, each outcome $h(Y_i)$ is weighed by the *likelihood ratio*

$$\frac{f(Y_i)}{g(Y_i)}. \quad - X \text{ density function}$$

In the preceding discussion, we have assumed that the random variable X has a density. The situation is similar when X is a discrete random variable. Suppose that the distribution of X is given by

$$\mathbb{P}(X = x_k) = p(x_k), \quad k = 1, 2, \ldots.$$

Consider a random variable Y with an alternative discrete probability distribution

$$\mathbb{P}(Y = x_k) = \bar{p}(x_k), \quad k = 1, 2, \ldots.$$

The importance sampling scheme generates iid copies of Y, say $\{Y_i\}$, and the estimate is the sample average of

$$h(Y_i) \frac{p(Y_i)}{\bar{p}(Y_i)}.$$

This estimate is unbiased as well.

Remark 7.1. To be more accurate, the alternative density function g should have the property that $f(x) = 0$ whenever $g(x) = 0$. In other words, the probability distribution associated with the density f is *absolutely continuous* with respect to the probability distribution associated with the density g. Analogous absolute continuity condition applies to the discrete probability distributions as well. Throughout the section, this requirement is implicitly imposed.

Remark 7.2. To ease exposition, we have assumed that X is a random variable. But the discussion obviously applies to the case where X is a general random vector, provided that one replaces the density function or the probability mass function by the joint density function or the joint probability mass function, respectively.

• what is the alternative sampling distribution?

7.1.1 Guideline for Selecting Alternative Distributions

The key question in importance sampling is the choice of the alternative sampling distribution. With a poor choice, the variance of the importance sampling estimate might even exceed that of the plain Monte Carlo estimate. In order to achieve the greatest variance reduction, we should determine the alternative sampling distribution that minimizes the variance of the importance sampling estimate.

To this end, let us assume that the random variable X has a density f and the importance sampling scheme generates iid samples $\{Y_i\}$ from an alternative density g. The variance of the importance sampling estimate is

$$\mathrm{Var}\left[\frac{1}{n}\sum_{i=1}^{n} h(Y_i)\frac{f(Y_i)}{g(Y_i)}\right] = \frac{1}{n}\mathrm{Var}\left[h(Y)\frac{f(Y)}{g(Y)}\right],$$

where Y is a representative sample with density g. Consider the special case where h is nonnegative. If one picks the alternative sampling density g to be proportional to $h(x)f(x)$, that is,

$$g(x) = ch(x)f(x)$$

for some constant c, then the importance sampling estimate has zero variance! To implement such an alternative density requires the knowledge of c. However,

$$1/c = \int_{\mathbb{R}} h(x)f(x)\,dx = E[h(X)] = \mu,$$

which is the very quantity we wish to estimate. Thus the zero variance estimate has no practical value. Nevertheless, this discussion is not pointless. It leads to the general guideline for selecting an alternative sampling distribution, which states that the density g should be chosen to "mimic" $h(x)f(x)$.

7.1.2 Importance Sampling for Normal Distributions

Importance sampling for normal distributions is of particular interest in financial engineering. For a normal distribution, the alternative sampling distribution is usually chosen from the class of normal distributions with the same variance. _(then how do you effect variance reduction)_

To be more concrete, consider the problem of estimating $E[h(X)]$ where X is a standard normal random variable. Let f be the density of $N(0,1)$. The alternative sampling distribution is chosen as $N(\theta, 1)$ for some $\theta \in \mathbb{R}$. Denote its density by $g(x)$. The corresponding likelihood ratio is

$$\frac{f(x)}{g(x)} = \exp\left\{-\theta x + \frac{1}{2}\theta^2\right\}.$$

In order to choose the parameter θ, recall that the principle is to pick a density $g(x)$ that mimics $h(x)f(x)$. This leads to the heuristic method of _mode matching_. That is, θ is chosen so as to match the mode of $h(x)f(x)$ with the mode of $g(x)$, which is precisely θ itself. In other words, we let $\theta = x^*$, where x^* maximizes $h(x)f(x)$, or equivalently,

$$\theta = \text{argmax}_x \, h(x)f(x).$$

The mode x^* can often be computed numerically. The above discussion easily extends to the multivariate normal distributions; see for example, Exercise 7.H.

Example 7.1. Assume that under the risk-neutral probability measure, the price of the underlying stock is a geometric Brownian motion

$$S_t = S_0 \exp\left\{\left(r - \frac{1}{2}\sigma^2\right)t + \sigma W_t\right\}.$$

Use importance sampling to estimate the price of a binary call option with maturity T and payoff

$$1_{\{S_T \geq K\}}.$$

SOLUTION: The plain Monte Carlo scheme is straightforward. We include the simulation results in Table 7.1. As a benchmark, the theoretical values are calculated from Example 2.3. The parameters are given by

$$S_0 = 50, \quad r = 0.01, \quad \sigma = 0.1, \quad T = 1, \quad n = 10000.$$

It is clear from the simulation results that as the strike price K becomes larger, the performance of the plain Monte Carlo estimate deteriorates. The reason is similar to that of Example 4.5. That is, with a moderate sample size, only a few or even none of the samples will reach the strike price K when K is large, which leads to poor estimates.

Table 7.1: Plain Monte Carlo for binary option

	$K = 50$	$K = 60$	$K = 70$	$K = 80$
Theoretical value	0.5148	0.0377	4.541×10^{-4}	1.643×10^{-6}
Estimate	0.5097	0.0392	6.000×10^{-4}	0
S.E.	0.0050	0.0019	2.449×10^{-4}	0
R.E.	0.98%	4.95%	40.81%	NaN

As for the importance sampling scheme, observe that the price of the option is

$$E\left[e^{-rT}1_{\{S_T \geq K\}}\right] = E\left[e^{-rT}1_{\{X \geq b\}}\right],$$

where $X = W_T/\sqrt{T}$ is a standard normal random variable and

$$b = \frac{\log(K/S_0) - (r - \sigma^2/2)T}{\sigma\sqrt{T}}.$$

In other words, the price of the option is $v = E[h(X)]$ where

$$h(x) = e^{-rT}1_{\{x \geq b\}}.$$

Let f be the density of the standard normal distribution. Then

$$h(x)f(x) = \begin{cases} 0 & \text{if } x < b, \\ e^{-rT}f(x) & \text{if } x \geq b, \end{cases}$$

and the mode of $h(x)f(x)$ is $x^* = \max\{b, 0\}$. The method of mode matching suggests $N(x^*, 1)$ as the alternative sampling distribution. Therefore, the importance sampling estimate is the sample average of iid copies of

$$h(Y)\frac{f(Y)}{g(Y)} = \begin{cases} 0 & \text{if } Y < b, \\ \exp\{-rT - x^*Y + (x^*)^2/2\} & \text{if } Y \geq b, \end{cases}$$

where Y is distributed as $N(x^*, 1)$.

Pseudocode for importance sampling:

> set $x^* = \max\{b, 0\}$
> for $i = 1, 2, \ldots, n$
>> generate a sample Y from $N(x^*, 1)$
>> set $H_i = 0$ if $Y < b$
>> set $H_i = \exp\{-rT - x^*Y + (x^*)^2/2\}$ if $Y \geq b$

compute the estimate $\hat{v} = \dfrac{1}{n} \sum_{i=1}^{n} H_i$

compute the standard error S.E. $= \sqrt{\dfrac{1}{n(n-1)} \left(\sum_{i=1}^{n} H_i^2 - n\hat{v}^2 \right)}.$

The simulation results are reported in Table 7.2. We use the same parameters and the sample size is again $n = 10000$.

Table 7.2: Importance sampling for binary call option

	$K = 50$	$K = 60$	$K = 70$	$K = 80$
Theoretical value	0.5148	0.0377	4.541×10^{-4}	1.643×10^{-6}
Estimate	0.5129	0.0379	4.717×10^{-4}	1.658×10^{-6}
S.E.	0.0050	0.0005	9.086×10^{-6}	3.796×10^{-8}
R.E.	0.97%	1.46%	1.93%	2.29%

The importance sampling scheme produces very accurate estimates even for large strike prices.   ∎

Example 7.2. The setup is the same as Example 7.1. Use importance sampling to estimate the price of a call option with maturity T and strike price K. Compare with the plain Monte Carlo scheme.

SOLUTION: The price of the call option is $v = E[h(X)]$, where X is a standard normal random variable and

$$h(x) = \left(S_0 \exp\left\{ -\frac{1}{2}\sigma^2 T + \sigma\sqrt{T}x \right\} - e^{-rT}K \right)^{+}.$$

We first run the plain Monte Carlo simulation and estimate the option price for a variety of strike prices. The parameters are

$$S_0 = 50, \ r = 0.05, \ \sigma = 0.2, \ T = 1, \ n = 10000.$$

The simulation results are reported in Table 7.3. The behavior of the plain Monte Carlo scheme is qualitatively similar to that of Example 7.1, that is, it performs poorly when the option is deep out of the money. The reason is the same—only for very few samples will the stock price at maturity exceed the strike price K when K is large. The theoretical values are obtained from the Black–Scholes formula.

Table 7.3: Plain Monte Carlo for call option

	$K = 50$	$K = 60$	$K = 80$	$K = 100$	$K = 120$
Theoretical value	5.2253	1.6237	0.0795	0.0024	6.066×10^{-5}
Estimate	5.2224	1.5943	0.0945	0.0035	0
S.E.	0.0740	0.0432	0.0104	0.0025	0
R.E.		1.42%	2.71%	11.00%	70.11%	NaN

Denote by f the density of $N(0, 1)$ and by g the alternative sampling density, which is assumed to be $N(\theta, 1)$. The mode matching method suggests that $\theta = x^*$, where x^* is the maximizer of $h(x)f(x)$. A typical picture of $h(x)f(x)$ is given in Figure 7.1.

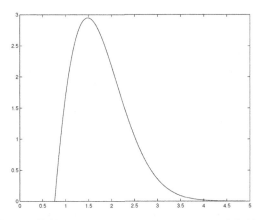

Figure 7.1: A representative picture of $h(x)f(x)$.

By taking the derivative of $h(x)f(x)$ and setting it to zero, it is easy to see that the maximizing x^* satisfies the equation

$$S_0 \exp\left\{-\frac{1}{2}\sigma^2 T + \sigma\sqrt{T}x^*\right\}(\sigma\sqrt{T} - x^*) + e^{-rT}Kx^* = 0, \qquad (7.1)$$

which can be solved numerically by the bisection method (see Remark 7.3).

Pseudocode for importance sampling:

solve for x^* from equation (7.1) by the bisection method

for $i = 1, 2, \ldots, n$

generate a sample Y from $N(x^*, 1)$

set $H_i = h(Y) \exp\{-x^*Y + (x^*)^2/2\}$

compute the estimate $\hat{v} = \dfrac{1}{n} \displaystyle\sum_{i=1}^{n} H_i$

compute the standard error S.E. $= \sqrt{\dfrac{1}{n(n-1)} \left(\displaystyle\sum_{i=1}^{n} H_i^2 - n\hat{v}^2 \right)}.$

The results from importance sampling are reported in Table 7.4. Compared with the plain Monte Carlo scheme, the importance sampling scheme yields much better estimates, especially when the strike price K is large.

Table 7.4: Importance sampling for call option

	$K = 50$	$K = 60$	$K = 80$	$K = 100$	$K = 120$
Theoretical value	5.2253	1.6237	0.0795	0.0024	6.066×10^{-5}
Estimate	5.2349	1.6310	0.0802	0.0024	5.957×10^{-5}
S.E.	0.0247	0.0112	0.0008	3.0×10^{-5}	8.631×10^{-7}
R.E.	0.47%	0.68%	1.00%	1.26%	1.45%
x^*	0.9849	1.4849	2.6002	3.6014	4.4569

Remark 7.3. The Bisection Method. Let h be a continuous function. The bisection method is a very simple algorithm for finding a root of the equation

$$h(x) = 0.$$

It is based on the intermediate value theorem, which asserts that if $h(x_1)$ and $h(x_2)$ have different signs, then there must exist a root on the interval $[x_1, x_2]$. To be more precise, suppose that $h(x_1) < 0 < h(x_2)$. Consider the midpoint

$$x_m = (x_1 + x_2)/2.$$

If $h(x_m) = 0$, then we have found a root. If $h(x_m) < 0$, then there must be a root on the interval $[x_m, x_2]$. Replace x_1 by x_m and repeat. Similarly, if $h(x_m) > 0$, then there must exist a root on $[x_1, x_m]$. Replace x_2 by x_m and repeat. Each iteration will halve the width of the interval, and the algorithm always converges to a root. The pseudocode for the bisection method is included. The algorithm will not end until a prescribed precision is achieved.

Pseudocode for solving $h(x) = 0$ by the bisection method:

initialize x_1 and x_2 — $h(x_1)$ and $h(x_2)$ must have opposite signs
prescribe a precision level ε
while $|x_1 - x_2| \geq \varepsilon$

$$\text{set } x_m = (x_1 + x_2)/2$$

if $h(x_m) = 0$, set $x_1 = x_m$ and $x_2 = x_m$

if $h(x_m) \neq 0$

set $x_1 = x_m$ if $h(x_m)$ has the same sign as $h(x_1)$

set $x_2 = x_m$ otherwise

return x_1. ∎

7.1.3 Importance Sampling for General Distributions

For a general random variable X, the alternative sampling distribution is often chosen from the so-called *exponential tilt* family. This is suggested by the asymptotic analysis of rare events under the right scaling [31, 33].

   a. X is continuous: Let f denote the density of X. The exponential tilt family consists of probability distributions with density

$$f_\theta(x) = \frac{1}{E[e^{\theta X}]} e^{\theta x} f(x)$$

   for some $\theta \in \mathbb{R}$.

   b. X is discrete: Assume that the probability mass function of X is p, that is, $\mathbb{P}(X = x) = p(x)$. The exponential tilt family consists of probability distributions with probability mass function

$$p_\theta(x) = \frac{1}{E[e^{\theta X}]} e^{\theta x} p(x)$$

   for some $\theta \in \mathbb{R}$.

The parameter θ is called the *tilting parameter*. The distribution determined by f_θ or p_θ is said to be the *exponential tilt distribution of X with parameter θ*. Note that when X is normally distributed, the exponential tilt family is the collection of all normal distributions with the same variance.

   It is beyond the scope of this book to state a general rule for selecting the best member from the exponential tilt family. But some of the ideas can be illustrated through the following example, where we estimate the probability of a large loss in a credit risk model. Importance sampling can be much more efficient than the plain Monte Carlo in this context since such probabilities are usually very small.

Example 7.3. Consider a much simplified credit risk model with m independent obligors. Denote by p_k the probability that the k-th obligor defaults and c_k the loss resulting from its default. Assuming that $c_k = 1$ for every k, the total loss is

$$L = \sum_{k=1}^{m} c_k X_k = \sum_{k=1}^{m} X_k,$$

where X_k is the default indicator for the k-th obligor, that is, $\{X_k\}$ are independent Bernoulli random variables such that

$$\mathbb{P}(X_k = 1) = p_k, \quad \mathbb{P}(X_k = 0) = 1 - p_k.$$

Assuming that x is a large threshold, use importance sampling to estimate the tail probability $\mathbb{P}(L > x)$.

SOLUTION: To exclude the trivial case, throughout the discussion we assume that

$$x > E[L] = \sum_{k=1}^{m} p_k.$$

Otherwise the event $\{L > x\}$ is not rare, and the plain Monte Carlo scheme suffices to produce an accurate estimate.

The alternative sampling distribution will be chosen from the exponential tilt family; see Exercise 7.7. Given $\theta \in \mathbb{R}$, let Y_k's be independent Bernoulli random variables such that

$$\bar{p}_k = \mathbb{P}(Y_k = 1) = \frac{1}{E[e^{\theta X_k}]} e^\theta \cdot \mathbb{P}(X_k = 1) = \frac{p_k e^\theta}{1 + p_k(e^\theta - 1)},$$

$$\mathbb{P}(Y_k = 0) = 1 - \bar{p}_k$$

for every k. Letting

$$\bar{L} = \sum_{k=1}^{m} Y_k,$$

the corresponding importance sampling estimate is the sample average of iid copies of

$$H = 1_{\{\bar{L}>x\}} \prod_{k=1}^{m} \left(\frac{p_k}{\bar{p}_k}\right)^{Y_k} \left(\frac{1 - p_k}{1 - \bar{p}_k}\right)^{1-Y_k}.$$

Since we would like to have $\bar{p}_k > p_k$ for every k so that it is more likely for the total loss to reach the threshold x under the alternative sampling distribution, θ is restricted to be positive.

It remains to choose an appropriate θ in order to achieve as much variance reduction as possible. Since H is an unbiased estimate, it suffices to make $E[H^2]$ as small as possible. Observe that

$$
\begin{aligned}
E[H^2] &= E\left[1_{\{\bar{L}>x\}}\prod_{k=1}^{m}\left(\frac{p_k}{\bar{p}_k}\right)^{2Y_k}\left(\frac{1-p_k}{1-\bar{p}_k}\right)^{2(1-Y_k)}\right] \\
&= E\left[1_{\{\bar{L}>x\}}e^{-2\theta\bar{L}+2\phi(\theta)}\right],
\end{aligned}
$$

where

$$
\phi(\theta) = \sum_{k=1}^{m}\log[1+p_k(e^{\theta}-1)].
$$

It follows that, for $\theta > 0$

$$
E[H^2] \le e^{-2\theta x + 2\phi(\theta)}. \tag{7.2}
$$

The idea is to find a θ that minimizes the upper bound (see Exercise 7.8), or equivalently,

$$
-2\theta x + 2\phi(\theta).
$$

It can be easily argued that the minimizing θ^* is the unique positive solution to the equation

$$
\phi'(\theta) = x \tag{7.3}
$$

when $x > E[L]$. The bisection method can be used to numerically solve for θ^*.

Pseudocode:

solve for the unique positive root θ^* from (7.3) by the bisection method
compute the corresponding $\bar{p}_k$ for $k = 1, \ldots, m$
for $i = 1, 2, \ldots, n$

generate Y_k from Bernoulli with parameter $\bar{p}_k$ for $k = 1, \ldots, m$
set $L = Y_1 + \cdots + Y_m$
set $H_i = \prod_{k=1}^{m}\left(\frac{p_k}{\bar{p}_k}\right)^{Y_k}\left(\frac{1-p_k}{1-\bar{p}_k}\right)^{1-Y_k}$ if $L > x$; otherwise set $H_i = 0$

compute the estimate $\hat{v} = \frac{1}{n}\sum_{i=1}^{n}H_i$

compute the standard error S.E. $= \sqrt{\frac{1}{n(n-1)}\left(\sum_{i=1}^{n}H_i^2 - n\hat{v}^2\right)}$.

The simulation results are presented in Table 7.5. The parameters are given by

$$m = 1000, \quad p_k = 0.01 \cdot [1 + e^{-k/m}], \quad k = 1, 2, \ldots, m.$$

The sample size is $n = 10000$.

Table 7.5: Credit risk model: Importance sampling versus plain Monte Carlo

	$x = 20$		$x = 30$		$x = 40$	
	IS	plain	IS	plain	IS	plain
Estimate	0.1482	0.1533	6.936×10^{-4}	0.0011	1.534×10^{-7}	0
S.E.	0.0019	0.0036	1.375×10^{-5}	0.0003	3.847×10^{-9}	0
R.E.	1.30%	2.35%	1.98%	30.14%	2.51%	NaN

Importance sampling proves to be much more efficient than the plain Monte Carlo scheme.   ∎

7.2   The Cross-Entropy Method

The cross-entropy method is a relatively new simulation technique originated by Reuven Rubinstein [28]. It is a very versatile methodology that can be employed to improve the efficiency of Monte Carlo simulation or solve difficult combinatorial optimization problems. We should mainly focus on the application of the cross-entropy method to importance sampling. A comprehensive treatment can be found in [29].

The cross-entropy method is essentially a simulation-based technique for selecting alternative sampling distributions. In order to illustrate the main idea, consider the generic problem of estimating the expected value

$$\mu = E[h(X)]$$

by importance sampling. To ease exposition, assume that X is a random variable with density $f(x)$ and h is a nonnegative function. The alternative sampling distribution is usually restricted to a parameterized family of density functions $\{f_\theta(x)\}$ that contains the original density f. A particularly popular choice is the exponential tilt family described in Section 7.1.3. For this reason, the reference parameter θ is often said to be the *tilting parameter* as well.

As we have discussed previously in Section 7.1.1, the importance sampling estimate based on the alternative sampling density

$$g^*(x) = \frac{1}{\mu}h(x)f(x) \tag{7.4}$$

has zero variance. Even though such a sampling distribution is impractical as it requires the knowledge of μ, it leads to the heuristic principle that an alternative sampling density "close" to $g^*(x)$ should be a good choice for importance sampling. The cross-entropy method aims to solve for the density $f_\theta(x)$ that is *closest* to $g^*(x)$ in the sense of *Kullback–Leibler cross entropy* or *relative entropy*, which is defined by

$$R(g^*\|f_\theta) = \int_{\mathbb{R}} \log \frac{g^*(x)}{f_\theta(x)} \cdot g^*(x)\,dx.$$

That is, the cross-entropy method chooses the minimizing density of the minimization problem

$$\min_\theta R(g^*\|f_\theta) \tag{7.5}$$

as the alternative sampling density for importance sampling. Plugging in formula (7.4), it follows that

$$
\begin{aligned}
R(g^*\|f_\theta) &= \int_{\mathbb{R}} g^*(x)\log g^*(x)\,dx - \int_{\mathbb{R}} g^*(x)\log f_\theta(x)\,dx \\
&= \int_{\mathbb{R}} g^*(x)\log g^*(x)\,dx - \frac{1}{\mu}\int_{\mathbb{R}} h(x)f(x)\log f_\theta(x)\,dx.
\end{aligned}
$$

Since neither the first term nor μ depends on θ, the minimization problem (7.5) is equivalent to the maximization problem

$$\max_\theta \int_{\mathbb{R}} h(x)f(x)\log f_\theta(x)\,dx. \tag{7.6}$$

This maximization problem does not admit explicit solutions in general. The cross-entropy method produces a simple, simulation-based, iterative algorithm to solve for the maximizing θ. At each step of the iteration, the maximizing θ is approximated by an *explicitly* computable quantity from a relatively small number of pilot samples. The extra computational cost incurred by these pilot samples and iterations is often significantly outweighed by the resulting variance reduction.

Remark 7.4. The Kullback–Leibler cross entropy is a measure of how close two probability distributions are. It is always nonnegative and takes value zero if and only if the two probability distributions coincide; see Exercise 7.10. Even though there are other definitions of distance between two probability distributions, the cross entropy is convenient as it is analytically simpler to work with.

Remark 7.5. Even though we have assumed that X is a continuous random variable for notational clarity, the cross-entropy method easily extends to the general cases where X is discrete or a random vector. Similarly, θ can also be a vector itself. All the ensuing discussions and formulas are still valid as long as one replaces the density functions with probability mass functions or their multivariate versions and replaces derivatives with gradients.

7.2.1   The Basic Cross-Entropy Algorithm

Consider the maximization problem (7.6). Since X has density f, one can rewrite

$$\int_{\mathbb{R}} h(x)f(x)\log f_\theta(x)\,dx = E\left[h(X)\log f_\theta(X)\right].$$

Under mild conditions such as $f_\theta(x)$ is differentiable with respect to θ and so on, the maximizer is the solution to the equation

$$0 = \frac{\partial}{\partial\theta}E[h(X)\log f_\theta(X)] = E\left[h(X)\frac{\partial}{\partial\theta}\log f_\theta(X)\right].$$

This equation is not explicitly solvable in general. However, if one replaces the expected value by sample average and considers the corresponding stochastic version

$$0 = \frac{1}{N}\sum_{k=1}^{N} h(X_k)\frac{\partial}{\partial\theta}\log f_\theta(X_k), \tag{7.7}$$

where $\{X_k\}$ are iid copies of X, then it is often possible to obtain a solution in *closed form*; see Exercise 7.11. Note that when θ is a vector, the partial derivative $\partial/\partial\theta$ denotes the gradient with respect to θ.

   The solution to (7.7) takes a particularly simple form in the context of normal distributions. More precisely, we have the following lemma, whose proof is straightforward and thus omitted.

Lemma 7.1. *Suppose that $f(x)$ is the density of $N(0, I_m)$ and $f_\theta(x)$ denotes the density of $N(\theta, I_m)$ where $\theta = (\theta_1, \ldots, \theta_m) \in \mathbb{R}^m$. Then the solution to (7.7) is simply*

$$\hat{\theta} = \frac{\sum_{k=1}^{N} h(X_k) X_k}{\sum_{k=1}^{N} h(X_k)},$$

where $\{X_1, \ldots, X_N\}$ are iid samples from $N(0, I_m)$.

To summarize, the cross-entropy method generates N iid pilot samples $X_1, \ldots, X_N$ from the original density $f(x)$ and compute $\hat{\theta}$ from equation (7.7). Once $\hat{\theta}$ is obtained, estimate $\mu = E[h(X)]$ by importance sampling with the alternative sampling density $f_{\hat{\theta}}(x)$. The pilot sample size N is usually chosen to be much smaller than the sample size for the estimation of μ. Below is the pseudocode.

Pseudocode for the basic cross-entropy algorithm:

generate N iid pilot samples $X_1, \ldots, X_N$ from density $f(x)$
obtain $\hat{\theta}$ by solving (7.7)
for $i = 1, 2, \ldots, n$

generate Y_i from the alternative sampling density $f_{\hat{\theta}}(x)$
set $H_i = h(Y_i) f(Y_i) / f_{\hat{\theta}}(Y_i)$.

compute the estimate $\hat{v} = \dfrac{1}{n} \sum_{i=1}^{n} H_i$

compute the standard error S.E. $= \sqrt{\dfrac{1}{n(n-1)} \left(\sum_{i=1}^{n} H_i^2 - n\hat{v}^2 \right)}$.

Example 7.4. Assume that under the risk-neutral probability measure, the underlying stock price is a geometric Brownian motion

$$S_t = S_0 \exp \left\{ \left(r - \frac{1}{2}\sigma^2 \right) t + \sigma W_t \right\},$$

where r is the risk-free interest rate. Estimate the price of a call option with strike price K and maturity T. Compare with Example 7.2.

SOLUTION: The price of the call option is $v = E[h(X)]$, where X is a standard normal random variable and

$$h(x) = \left(S_0 \exp \left\{ -\frac{1}{2}\sigma^2 T + \sigma\sqrt{T}x \right\} - e^{-rT} K \right)^+.$$

The family of alternative sampling densities is $\{f_\theta : \theta \in \mathbb{R}\}$, where f_θ is the density of $N(\theta, 1)$. The solution $\hat{\theta}$ to equation (7.7) is given by Lemma 7.1. Below is the pseudocode.

Pseudocode for call option by the cross-entropy method:

> generate N iid pilot samples $X_1, \ldots, X_N$ from $N(0,1)$
> set $\hat{\theta} = \sum_{k=1}^{N} h(X_k) X_k / \sum_{k=1}^{N} h(X_k)$
> for $i = 1, 2, \ldots, n$
>
> > generate a sample Y from $N(\hat{\theta}, 1)$
> > set $H_i = h(Y) \exp\{-\hat{\theta} Y + \hat{\theta}^2/2\}$
>
> compute the estimate $\hat{v} = \dfrac{1}{n} \sum_{i=1}^{n} H_i$
>
> compute the standard error S.E. $= \sqrt{\dfrac{1}{n(n-1)} \left(\sum_{i=1}^{n} H_i^2 - n\hat{v}^2 \right)}$.

The numerical results are reported in Table 7.6. The pilot sample size is $N = 2000$, and the sample size for importance sampling is $n = 10000$. The parameters are again given by

$$S_0 = 50, \quad r = 0.05, \quad \sigma = 0.2, \quad T = 1.$$

Table 7.6: Basic cross-entropy method for call option

	$K=50$	$K=60$	$K=80$	$K=100$		$K=120$
True value	5.2253	1.6237	0.0795	0.0024		6.066×10^{-5}
Estimate	5.2166	1.6273	0.0805	0.0024	NaN	NaN
S.E.	0.0243	0.0109	0.0008	3.2×10^{-5}	NaN	NaN
R.E.	0.47%	0.67%	0.99%	1.32%	NaN	NaN
$\hat{\theta}$	1.2057	1.7525	2.8247	3.3781	NaN	NaN

Compared with the simulation results in Example 7.2, the performance of the basic cross-entropy scheme is indistinguishable from that of the importance sampling scheme using the mode matching method, when K is not exceedingly large. However, around $K = 100$, the basic cross-entropy scheme starts to behave erratically: sometimes it yields a great estimate, and sometimes it produces an "NaN", which stands for "Not a Number" in MATLAB®. When $K = 120$, almost all the results from the cross-entropy method are "NaN". It is because in the estimation of $\hat{\theta}$, the denominator

$$\sum_{k=1}^{N} h(X_k)$$

is more likely to become zero as K increases. This example suggests that a modification of the basic cross-entropy algorithm is necessary when the simulation is dominated by events with very small probabilities. Such development will be discussed later in the book. ■

Example 7.5. Consider a discretely monitored average price call option with payoff $(\bar{S} - K)^+$ and maturity T. Here $\bar{S}$ is the arithmetic mean

$$\bar{S} = \frac{1}{m} \sum_{k=1}^{m} S_{t_k}$$

for a given set of dates $0 < t_1 < \cdots < t_m = T$. Assume that under the risk-neutral probability measure the price of the underlying asset is a geometric Brownian motion

$$S_t = S_0 \exp \left\{ \left(r - \frac{1}{2}\sigma^2 \right) t + \sigma W_t \right\},$$

where r is the risk-free interest rate. Estimate the option price.

SOLUTION: For $1 \le k \le m$, let $Z_k = (W_{t_k} - W_{t_{k-1}})/\sqrt{t_k - t_{k-1}}$. Then $Z_1, \ldots, Z_m$ are iid standard normal random variables and $X = (Z_1, \ldots, Z_m)$ is an m-dimensional standard normal random vector. The stock prices at time t_k can be written as a function of X:

$$S_{t_k} = S_0 \exp \left\{ \left(r - \frac{1}{2}\sigma^2 \right) t_k + \sigma \sum_{j=1}^{k} \sqrt{t_j - t_{j-1}} Z_j \right\}. \tag{7.8}$$

Consequently, the discounted option payoff can also be written as a function of X, say

$$h(X) = e^{-rT} (\bar{S} - K)^+. \tag{7.9}$$

Denote the joint density function of X by f, that is, for $x = (x_1, \ldots, x_m)$,

$$f(x) = \left(\frac{1}{\sqrt{2\pi}} \right)^m \exp \left\{ -\frac{1}{2} \sum_{i=1}^{m} x_i^2 \right\}.$$

Suppose that the family of alternative sampling densities is $\{N(\theta, I_m) : \theta \in \mathbb{R}^m\}$. Let f_θ be the density of $N(\theta, I_m)$, that is,

$$f_\theta(x) = \left(\frac{1}{\sqrt{2\pi}}\right)^m \exp\left\{-\frac{1}{2}\sum_{i=1}^{m}(x_i - \theta_i)^2\right\}$$

for $\theta = (\theta_1, \ldots, \theta_m)$. The importance sampling estimate is the sample average of iid copies of

$$h(Y)\frac{f(Y)}{f_\theta(Y)} = h(Y)\exp\left\{-\sum_{i=1}^{m}\theta_i Y_i + \frac{1}{2}\sum_{i=1}^{m}\theta_i^2\right\},$$

where $Y = (Y_1, \ldots, Y_m)$ has distribution $N(\theta, I_m)$, that is, $\{Y_1, \ldots, Y_m\}$ are independent and Y_i is distributed as $N(\theta_i, 1)$ for each i.

A good tilting parameter θ can be determined by the basic cross-entropy scheme. Indeed, let $X_1, \ldots, X_N$ be iid pilot samples from the original distribution $N(0, I_m)$. Then the solution to equation (7.7) is given by Lemma 7.1, i.e.,

$$\hat{\theta} = \frac{\sum_{k=1}^{N} h(X_k) X_k}{\sum_{k=1}^{N} h(X_k)}.$$

Below is the pseudocode.

Pseudocode for average price call by the cross-entropy method:

    generate iid pilot samples $X_1, \ldots, X_N$ from $N(0, I_m)$

    set $\hat{\theta} = (\hat{\theta}_1, \ldots, \hat{\theta}_m) = \sum_{k=1}^{N} h(X_k) X_k / \sum_{k=1}^{N} h(X_k)$

($\triangleright$)   for $i = 1, 2, \ldots, n$

        for $j = 1, 2, \ldots, m$

            generate Y_j from $N(\hat{\theta}_j, 1)$

            set $S_{t_j} = S_{t_{j-1}} \exp\left\{(r - \sigma^2/2)(t_j - t_{j-1}) + \sigma\sqrt{t_j - t_{j-1}}Y_j\right\}$

($\square$)   compute the discounted payoff multiplied by the likelihood ratio

$$H_i = e^{-rT}(\bar{S} - K)^+ \cdot \exp\left\{-\sum_{j=1}^{m}\hat{\theta}_j Y_j + \frac{1}{2}\sum_{j=1}^{m}\hat{\theta}_j^2\right\}$$

    compute the estimate $\hat{v} = \frac{1}{n}\sum_{i=1}^{n}H_i$

    compute the standard error S.E. $= \sqrt{\frac{1}{n(n-1)}\left(\sum_{i=1}^{n}H_i^2 - n\hat{v}^2\right)}.$

We should mention that the first two lines of the pseudocode calculate the tilting parameter using the cross-entropy method. The details of evaluating $h(X_k)$ are left out for two reasons: (i) it is straightforward from (7.9), the definition of h; (ii) it is indeed very much similar to the lines from ($\triangleright$) to ($\square$). We will fill in the missing details below for the purpose of illustration. However, such details will not be given for the rest of the book in order to avoid repetition.

More details for the first two lines in the previous pseudocode:

for $k = 1, 2, \ldots, N$

   for $j = 1, 2, \ldots, m$

      generate Z_j from $N(0,1)$

      set $S_{t_j} = S_{t_{j-1}} \exp \left\{ (r - \sigma^2/2)(t_j - t_{j-1}) + \sigma \sqrt{t_j - t_{j-1}} Z_j \right\}$

   set $X_k = (Z_1, \ldots, Z_m)$

   set $R_k = h(X_k) = e^{-rT} (\bar{S} - K)^+$

set $\hat{\theta} = (\hat{\theta}_1, \ldots, \hat{\theta}_m) = \sum_{k=1}^N R_k X_k / \sum_{k=1}^N R_k$

The numerical results are reported in Table 7.7. We compare the cross-entropy method with the plain Monte Carlo for

$$S_0 = 50, \quad r = 0.05, \quad \sigma = 0.2, \quad T = 1, \quad m = 12, \quad t_i = \frac{i}{m}.$$

The sample size for importance sampling is $n = 10000$ and the pilot sample size for the cross-entropy method is $N = 2000$. The cross-entropy method reduces the variance significantly, especially when K is large.

Table 7.7: Cross-entropy method versus plain Monte Carlo

Strike price	$K = 50$		$K = 60$		$K = 70$	
	CE	plain MC	CE	plain MC	CE	plain MC
Estimate	3.0637	3.0998	0.3404	0.3178	0.0168	0.0237
S.E.	0.0155	0.0429	0.0033	0.0144	0.0004	0.0042
R.E.	0.51%	1.38%	0.96%	4.54%	2.44%	17.8%

It should be noted that the mode matching method can also be applied to this problem. Even though there are m unknowns, it can be reduced to a single equation, which can then be solved by the bisection method [11]. ∎

7.2.2   The General Iterative Cross Entropy Algorithm

The basic cross-entropy algorithm is the simple version of a more general iterative procedure for solving the maximization problem (7.6). The latter is particularly useful in the context of simulating events with very small probabilities.

Every iteration in this general scheme involves two phases. In the i-th iteration, one first generates iid samples from the density function $f_\theta(x)$ with $\theta = \hat{\theta}^{i-1}$ being the current candidate of the tilting parameter. The tilting parameter $\hat{\theta}^{i-1}$ is then updated to $\hat{\theta}^i$ based on these samples. Here we have used a superscript instead of a subscript because it is very common that θ is a vector and we would like to reserve the subscript for the individual components of θ. As in the basic algorithm, $\hat{\theta}^i$ often admits *analytical* formulas. If $\hat{\theta}^*$ is the tilting parameter from the final iteration, then $f_{\hat{\theta}^*}$ is used as the alternative sampling density in the importance sampling scheme to estimate $\mu = E[h(X)]$, the quantity of interest.

The iterative cross-entropy algorithm is based on the following observation. Define the likelihood ratio

$$\ell_\theta(x) = \frac{f(x)}{f_\theta(x)}. \tag{7.10}$$

Consider the maximization problem (7.6). Fixing an *arbitrary* tilting parameter, say v, we can rewrite the integral in (7.6) as

$$\int_{\mathbb{R}} h(x)\ell_v(x) \log f_\theta(x) \cdot f_v(x)\, dx = E\left[h(Y)\ell_v(Y) \log f_\theta(Y)\right],$$

where Y is a random variable with density $f_v(x)$. Consequently, the maximizer to (7.6) satisfies the equation

$$0 = E\left[h(Y)\ell_v(Y)\frac{\partial}{\partial\theta} \log f_\theta(Y)\right]. \tag{7.11}$$

As before, we replace the expected value by sample average and solve the equation

$$0 = \frac{1}{N}\sum_{k=1}^{N} h(Y_k)\ell_v(Y_k)\frac{\partial}{\partial\theta} \log f_\theta(Y_k),$$

where $Y_1, \ldots, Y_N$ are iid pilot samples from the density f_v. This leads to the following updating rule for $\hat{\theta}$:

The updating rule of $\hat{\theta}$. Suppose that $\hat{\theta}^j$ is the value of the tilting parameter at the end of the j-th iteration. Let $\hat{\theta}^{j+1}$ be the solution to the equation

$$0 = \frac{1}{N} \sum_{k=1}^{N} h(Y_k)\ell_{\hat{\theta}^j}(Y_k)\frac{\partial}{\partial\theta}\log f_\theta(Y_k), \qquad (7.12)$$

where $Y_1, \ldots, Y_N$ are iid pilot samples from the density $f_{\hat{\theta}^j}$.

Equation (7.12) is of exactly the same form as the basic cross-entropy equation (7.7), except that X_k is replaced by Y_k and $h(X_k)$ by $h(Y_k)\ell_{\hat{\theta}^j}(Y_k)$. Therefore, just like (7.7) it is often explicitly solvable; see Exercise 7.11. In particular, we have the following result, which is an immediate corollary of Lemma 7.1.

Lemma 7.2. *Suppose that $f(x)$ is the density of $N(0, I_m)$ and $f_\theta(x)$ is the density of $N(\theta, I_m)$ where $\theta = (\theta_1, \ldots, \theta_m) \in \mathbb{R}^m$. Then the solution to (7.12) is simply*

$$\hat{\theta}^{j+1} = \frac{\sum_{k=1}^{N} h(Y_k)\ell_{\hat{\theta}^j}(Y_k)Y_k}{\sum_{k=1}^{N} h(Y_k)\ell_{\hat{\theta}^j}(Y_k)} = \frac{\sum_{k=1}^{N} h(Y_k)e^{-\langle\hat{\theta}^j, Y_k\rangle}Y_k}{\sum_{k=1}^{N} h(Y_k)e^{-\langle\hat{\theta}^j, Y_k\rangle}},$$

where $\{Y_1, \ldots, Y_N\}$ are iid samples from $N(\hat{\theta}^j, I_m)$ and $\langle\cdot, \cdot\rangle$ is the inner product of two vectors defined by

$$\langle\theta, y\rangle = \sum_{i=1}^{m} \theta_i y_i, \quad \text{if } \theta = (\theta_1, \ldots, \theta_m), \, y = (y_1, \ldots, y_m).$$

Note that in the one-dimensional case (i.e., $m = 1$), the inner product is just the regular product of numbers.

If the initial tilting parameter is reasonably chosen, the number of iterations necessary to reach a good final tilting parameter is very small in practice: four or five iterations are in general sufficient, and for many problems one or two iterations are all that is needed. Since $\hat{\theta}^j$'s are estimates based on samples, we cannot expect the convergence in the classical sense that $|\hat{\theta}^{j+1} - \hat{\theta}^j| \to 0$ as j tends to infinity. The "convergence" is reached when $\hat{\theta}^j$ starts to oscillate with small variations.

The choice of the initial tilting parameter $\hat{\theta}^0$ is quite flexible in general. For example, in many situations it suffices to let $\hat{\theta}^0 = \theta^0$, where θ^0 corresponds to the original distribution $f(x)$, that is, $f = f_{\theta^0}$. With this initial choice, $\ell_{\hat{\theta}^0}(x) = 1$ and equation (7.12) reduces to (7.7). In other words, the basic cross-entropy algorithm is equivalent to the *first* iteration of the general iterative scheme with $\hat{\theta}^0 = \theta^0$. However, when the simulation is related to events with very small probabilities, the choice of $\hat{\theta}^0$ is less straightforward. The reason is as follows. As we have observed in Example 7.4, when the simulation is largely determined by rare events, the basic cross-entropy algorithm, and hence the first iteration of the general scheme, will frequently produce "NaN" if one lets $\hat{\theta}^0 = \theta^0$.

In financial engineering, many quantities of interest are the expected values of random variables that can only be nonzero on sets of the form $\{R \geq a\}$. The random variable R can be, for example, the underlying asset price or the total loss in a risk model, and a can be some given level related to the strike price, barrier, or loss threshold, and so on. The difficulty arises when a is large. The rule of thumb for a good initial tilting parameter $\hat{\theta}^0$ is that it should be chosen in an "economical" way, or equivalently, the original distribution should be tilted just enough, to ensure that $\{R \geq a\}$ is no longer rare. This in general means that we should let

$$E_{\hat{\theta}^0}[R] = a,$$

where $E_{\hat{\theta}^0}[\cdot]$ denotes the expected value taken under the alternative sampling density $f_{\hat{\theta}^0}$. This expected value may or may not admit an analytical expression, but it can often be approximated.

We should demonstrate how to judiciously choose an initial tilting parameter through examples in this section and postpone the discussion of a general initialization technique to the next.

Example 7.6. Let us revisit the problem of estimating the call option price in Example 7.4. The basic cross-entropy method fails when the strike price K is exceedingly large. Design an iterative cross-entropy scheme to resolve this issue.

SOLUTION: Recall that the call option price is $v = E[h(X)]$, where X is a standard normal random variable and

$$h(X) = \left(S_0 \exp\left\{ -\frac{1}{2}\sigma^2 T + \sigma\sqrt{T}X \right\} - e^{-rT}K \right)^+ .$$

The family of alternative sampling densities is $\{ f_\theta : \theta \in \mathbb{R} \}$, where f_θ is the density of $N(\theta, 1)$. The updating rule of $\hat{\theta}^j$ is given by Lemma 7.2 and is straightforward. The key issue is the initialization of the tilting parameter, especially when K is large.

A good initial tilting parameter $\hat{\theta}^0$ should alter the original distribution just enough so that $\{h(Y) > 0\}$ is no longer a rare event if Y is distributed as $N(\hat{\theta}^0, 1)$, or equivalently, the stock price exceeds K with nontrivial probability under the alternative sampling density $f_{\hat{\theta}^0}$. Under these considerations, a natural choice of $\hat{\theta}^0$ is such that

$$E\left(S_0 \exp\left\{-\frac{1}{2}\sigma^2 T + \sigma\sqrt{T}Y\right\} - e^{-rT}K\right) = 0,$$

where Y is normally distributed as $N(\hat{\theta}^0, 1)$. It is straightforward to calculate the expected value and solve the equation to obtain

$$\hat{\theta}^0 = \frac{1}{\sigma\sqrt{T}}\log\left(\frac{K}{S_0}\right) - \frac{r}{\sigma}\sqrt{T}. \tag{7.13}$$

Below is the pseudocode. The total number of iterations is denoted by IT_NUM.

> **Pseudocode for call option by the iterative cross-entropy method:**
>
> initialize $\hat{\theta}^0$ by (7.13) and set the iteration counter $j = 0$
> (*) generate N iid samples $Y_1, \ldots, Y_N$ from $N(\hat{\theta}^j, 1)$
> set $\hat{\theta}^{j+1} = \sum_{k=1}^{N} h(Y_k)e^{-\hat{\theta}^j Y_k}Y_k / \sum_{k=1}^{N} h(Y_k)e^{-\hat{\theta}^j Y_k}$
> set the iteration counter $j = j + 1$
> if $j = $ IT_NUM set $\hat{\theta} = \hat{\theta}^j$ and continue, otherwise go to step (*)
> for $i = 1, 2, \ldots, n$
>
> > generate a sample Y from $N(\hat{\theta}, 1)$
> > set $H_i = h(Y)\exp\{-\hat{\theta}Y + \hat{\theta}^2/2\}$
>
> compute the estimate $\hat{v} = \frac{1}{n}\sum_{i=1}^{n} H_i$
>
> compute the standard error S.E. $= \sqrt{\frac{1}{n(n-1)}\left(\sum_{i=1}^{n} H_i^2 - n\hat{v}^2\right)}.$

The numerical results are reported in Table 7.8. The pilot sample size is $N = 2000$ and the sample size for importance sampling is $n = 10000$. The number of iteration is IT_NUM $= 5$. The parameters are again given by

$$S_0 = 50, \quad r = 0.05, \quad \sigma = 0.2, \quad T = 1.$$

Compared with the simulation results of the basic cross-entropy algorithm in Example 7.2, the improvement from the general iterative cross-entropy method is negligible when K is moderate. The reason is that if one regards the basic cross-entropy algorithm as a special case of the general iterative scheme with $\hat{\theta}^0 = \theta^0 = 0$ and IT_NUM $= 1$, then the convergence is already reached with one iteration. On the other hand, the general scheme consistently yields very accurate estimates when K is large.

Table 7.8: Iterative cross-entropy method for call option

	$K = 50$	$K = 60$	$K = 80$	$K = 100$	$K = 120$
True value	5.2253	1.6237	0.0795	0.0024	6.066×10^{-5}
Estimate	5.2603	1.6338	0.0791	0.0024	6.042×10^{-5}
S.E.	0.0240	0.0109	0.0008	2.987×10^{-5}	8.667×10^{-7}
R.E.	0.46%	0.67%	1.01%	1.25%	1.43%
$\hat{\theta}$	1.2330	1.7632	2.8631	3.8209	4.6657

Table 7.9: Iteration of the tilting parameter $\hat{\theta}^j$

	$\hat{\theta}^0$	$\hat{\theta}^1$	$\hat{\theta}^2$	$\hat{\theta}^3$	$\hat{\theta}^4$	$\hat{\theta}^5 = \hat{\theta}$
$K = 50$	-0.2500	1.1770	1.2127	1.2381	1.2092	1.2330
$K = 60$	0.6616	1.7681	1.7468	1.7802	1.7814	1.7632
$K = 80$	2.1000	2.8451	2.8605	2.8731	2.8676	2.8631
$K = 100$	3.2157	3.8165	3.8223	3.8196	3.8162	3.8209
$K = 120$	4.1273	4.6518	4.6508	4.6479	4.6472	4.6657

The iteration of $\hat{\theta}^j$ is given in Table 7.9. We can see that the convergence is achieved within one or two iterations for all strike prices.   ■

Example 7.7. Revisit Example 7.5 for the pricing of average price call options. When the strike price K gets larger, the plain Monte Carlo scheme or the basic cross-entropy method will start to fail. Design an iterative cross-entropy scheme to estimate the option price.

SOLUTION: Recall that the discounted option payoff is denoted by $h(X)$, where $X = (Z_1, \ldots, Z_m)$ and

$$Z_k = (W_{t_k} - W_{t_{k-1}})/\sqrt{t_k - t_{k-1}}, \quad 1 \le k \le m.$$

The joint density function of X and the family of alternative sampling densities are given by $f(x)$ and $\{f_\theta(x)\}$, respectively. Furthermore, for

$x = (x_1, \ldots, x_m)$ and $\theta = (\theta_1, \ldots, \theta_m)$, the likelihood ratio $\ell_\theta(x)$ is given by

$$\ell_\theta(x) = \frac{f(x)}{f_\theta(x)} = \exp\left\{-\sum_{i=1}^{m} x_i\theta_i + \frac{1}{2}\sum_{i=1}^{m}\theta_i^2\right\}.$$

In the general iterative cross-entropy scheme, the updating rule for the tilting parameter is given by Lemma 7.2:

$$\hat{\theta}^{j+1} = \frac{\sum_{k=1}^{N} h(Y_k)\ell_{\hat{\theta}^j}(Y_k)Y_k}{\sum_{k=1}^{N} h(Y_k)\ell_{\hat{\theta}^j}(Y_k)} = \frac{\sum_{k=1}^{N} h(Y_k)e^{-\langle \hat{\theta}^j, Y_k\rangle}Y_k}{\sum_{k=1}^{N} h(Y_k)e^{-\langle \hat{\theta}^j, Y_k\rangle}},$$

where Y_k's are iid samples from $f_{\hat{\theta}^j}$ or $N(\hat{\theta}^j, I_m)$.

A good choice of $\hat{\theta}^0$ should tilt the original distribution just enough so that $\{h(Y) > 0\}$ happens with moderate probability if Y is distributed as $N(\hat{\theta}^0, I_m)$, or equivalently, $\bar{S}$ exceeds K with nontrivial probability if $(Z_1, \ldots, Z_m)$ is treated as a $N(\hat{\theta}^0, I_m)$ random vector, in which case it follows from (7.8) that

$$E[\bar{S}] = \frac{1}{m}\sum_{k=1}^{m} E[S_{t_k}] = \frac{1}{m}\sum_{k=1}^{m} S_0 \exp\left\{rt_k + \sigma\sum_{i=1}^{k}\sqrt{t_i - t_{i-1}}\,\hat{\theta}_i^0\right\}.$$

A convenient choice is to set $\hat{\theta}^0 = (a, \ldots, a)$ and let $E[\bar{S}] = K$, that is, a satisfies

$$\frac{1}{m}\sum_{k=1}^{m} S_0 \exp\left\{rt_k + a\sigma\sum_{i=1}^{k}\sqrt{t_i - t_{i-1}}\right\} = K. \tag{7.14}$$

This equation can be easily solved by the bisection method. Below is the pseudocode for the iterative cross-entropy scheme. The number of iterations is denoted by IT_NUM.

Pseudocode for average price call by the iterative cross-entropy method:

> Solve for a from equation (7.14) by the bisection method
>
> initialize $\hat{\theta}^0 = (a, \ldots, a)$ and set the iteration counter $j = 0$
>
> (*) generate N iid samples $Y_1, \ldots, Y_N$ from $N(\hat{\theta}^j, I_m)$
>
> set $\hat{\theta}^{j+1} = \sum_{k=1}^{N} h(Y_k)e^{-\langle \hat{\theta}^j, Y_k\rangle}Y_k / \sum_{k=1}^{N} h(Y_k)e^{-\langle \hat{\theta}^j, Y_k\rangle}$
>
> set the iteration counter $j = j + 1$
>
> if $j = $ IT_NUM set $\hat{\theta} = \hat{\theta}^j$ and continue, otherwise go to step (*)
>
> for $i = 1, 2, \ldots, n$

for $k = 1, 2, \ldots, m$

    generate Y_k from $N(\hat{\theta}_k, 1)$

    set $S_{t_k} = S_{t_{k-1}} \exp \left\{ (r - \sigma^2/2)(t_k - t_{k-1}) + \sigma\sqrt{t_k - t_{k-1}} Y_k \right\}$

compute the discounted payoff multiplied by the likelihood ratio

$$H_i = e^{-rT}(\bar{S} - K)^+ \cdot \exp \left\{ -\sum_{k=1}^{m} \hat{\theta}_k Y_k + \frac{1}{2} \sum_{k=1}^{m} \hat{\theta}_k^2 \right\}$$

compute the estimate $\hat{v} = \dfrac{1}{n} \displaystyle\sum_{i=1}^{n} H_i$

compute the standard error S.E. $= \sqrt{\dfrac{1}{n(n-1)} \left(\displaystyle\sum_{i=1}^{n} H_i^2 - n\hat{v}^2 \right)}.$

The numerical results are reported in Table 7.10. The parameters are again given by

$$S_0 = 50, \quad r = 0.05, \quad \sigma = 0.2, \quad T = 1, \quad m = 12, \quad t_i = \frac{i}{m}.$$

The sample size is always $n = 10000$ for importance sampling. The pilot sample size for the cross-entropy method is $N = 2000$ and the number of iterations is IT_NUM $= 5$.

Table 7.10: Iterative cross-entropy scheme for Asian option

	$K = 50$	$K = 60$	$K = 70$	$K = 80$	$K = 100$
Estimate	3.0749	0.3413	0.0166	4.983×10^{-4}	2.083×10^{-7}
S.E.	0.0151	0.0028	0.0002	6.538×10^{-6}	3.389×10^{-9}
R.E.	0.49%	0.81%	1.11%	1.31%	1.63%

The performance of the iterative cross-entropy scheme is indistinguishable from that of the basic cross-entropy scheme when K is not very large. When K reaches 80 and above, the basic cross-entropy scheme frequently gives an "NaN", while the iterative scheme consistently yields accurate estimates. ∎

Example 7.8. Consider a multi-asset basket call option with maturity T and payoff

$$\left(c_1 S_T^{(1)} + \cdots + c_d S_T^{(d)} - K \right)^+.$$

Under the risk-neutral probability measure, the underlying stock prices are assumed to be geometric Brownian motions:

$$S_t^{(i)} = S_0^{(i)} \left\{ \left(r - \frac{1}{2}\sigma_i^2 \right) t + \sigma_i W_t^{(i)} \right\}, \quad i = 1, \ldots, d,$$

where $W = (W^{(1)}, \ldots, W^{(d)})$ is a d-dimensional Brownian motion with co-variance matrix $\Sigma = [\Sigma_{ij}]$ such that $\Sigma_{ii} = 1$ for all i. Use the cross-entropy method to estimate the option price.

SOLUTION: Let A be a Cholesky factorization of Σ, that is, A is a lower triangular matrix with $AA' = \Sigma$. It follows that

$$\frac{1}{\sqrt{T}} W_T = \frac{1}{\sqrt{T}} (W_T^{(1)}, \ldots, W_T^{(d)}) = AX \tag{7.15}$$

for some d-dimensional standard normal random vector $X = (Z_1, \ldots, Z_d)'$. Clearly, the discounted option payoff can be written as $h(AX)$, where

$$h(y) = \left(\sum_{i=1}^{d} c_i S_0^{(i)} \left\{ -\frac{1}{2}\sigma_i^2 T + \sigma_i \sqrt{T} y_i \right\} - e^{-rT} K \right)^+$$

for $y = (y_1, \ldots, y_d)$. Except for the payoff function, the setup is very similar to Example 7.7. Let $f(x)$ be the joint density function of X or $N(0, I_d)$. The family of alternative sampling densities is given by $\{ f_\theta(x) : \theta \in \mathbb{R}^d \}$, where f_θ is the joint density function for $N(\theta, I_d)$. The updating rule in the iterative cross-entropy scheme is again

$$\hat{\theta}^{j+1} = \frac{\sum_{k=1}^{N} h(Y_k) e^{-\langle \hat{\theta}^j, Y_k \rangle} Y_k}{\sum_{k=1}^{N} h(Y_k) e^{-\langle \hat{\theta}^j, Y_k \rangle}},$$

where $\hat{\theta}^j$ denotes the value of the tilting parameter at the end of the j-th iteration and Y_k's are iid samples from $N(\hat{\theta}^j, I_d)$.

To initialize the tilting parameter, observe that if X were $N(\theta, I_d)$, then it follows from (7.15) that $W_T/\sqrt{T}$ would have distribution $N(A\theta, \Sigma)$. In particular, $W_T^{(i)}/\sqrt{T}$ would be distributed as $N(\eta_i, 1)$, where $\eta = (\eta_1, \ldots, \eta_d)' = A\theta$, which implies that

$$E \left[\sum_{i=1}^{d} c_i S_T^{(i)} \right] = \sum_{i=1}^{d} c_i S_0^{(i)} \exp \left\{ rT + \sigma_i \sqrt{T} \eta_i \right\}.$$

Therefore, a convenient choice will be to choose $\eta = (a, \ldots, a)'$ so that the above expected value equals K, or

$$\sum_{i=1}^{d} c_i S_0^{(i)} \exp\left\{ rT + \sigma_i \sqrt{T}a \right\} = K, \tag{7.16}$$

and then set $\hat{\theta}^0 = A^{-1}\eta$. The solution to equation (7.16) can be easily obtained by the bisection method. Below is the pseudocode.

Pseudocode for basket call by the iterative cross-entropy method:

> solve for a from equation (7.16) by the bisection method
> set $\eta = (a, \ldots, a)'$
> initialize $\hat{\theta}^0 = A^{-1}\eta$ and set the iteration counter $j = 0$
> (*) generate N iid samples $Y_1, \ldots, Y_N$ from $N(\hat{\theta}^j, I_d)$
> set $\hat{\theta}^{j+1} = \sum_{k=1}^{N} h(AY_k)e^{-\langle \hat{\theta}^j, Y_k \rangle}Y_k / \sum_{k=1}^{N} h(AY_k)e^{-\langle \hat{\theta}^j, Y_k \rangle}$
> set the iteration counter $j = j + 1$
> if $j = $ IT_NUM set $\hat{\theta} = \hat{\theta}^j$ and continue, otherwise go to step (*)
> for $i = 1, 2, \ldots, n$
> > generate $Y = (Y_1, \ldots, Y_d)$ from $N(\hat{\theta}, I_d)$
> > compute the discounted payoff multiplied by the likelihood ratio
> >
> > $$H_i = h(AY) \cdot \exp\left\{ -\sum_{k=1}^{m} \hat{\theta}_k Y_k + \frac{1}{2} \sum_{k=1}^{m} \hat{\theta}_k^2 \right\}$$
> >
> > compute the estimate $\hat{v} = \frac{1}{n}\sum_{i=1}^{n} H_i$
> >
> > compute the standard error S.E. $= \sqrt{\frac{1}{n(n-1)}\left(\sum_{i=1}^{n} H_i^2 - n\hat{v}^2 \right)}.$

We consider a numerical example with $d = 4$ underlying assets. The simulation results are shown in Table 7.11 for

$$S_0^{(1)} = 45, \ S_0^{(2)} = S_0^{(3)} = 50, \ S_0^{(4)} = 55, \ r = 0.03, \ T = 0.5, \ c_1 = 0.4,$$

$$c_2 = c_3 = c_4 = 0.2, \ \sigma_1 = \sigma_2 = 0.1, \ \sigma_3 = \sigma_4 = 0.2,$$

$$\Sigma = \begin{bmatrix} 1.0 & 0.5 & -0.3 & 0.4 \\ 0.5 & 1.0 & 0.3 & 0.5 \\ -0.3 & 0.3 & 1.0 & 0.7 \\ 0.4 & 0.5 & 0.7 & 1.0 \end{bmatrix}, \ n = 10000.$$

We use IT_NUM $= 5$ iterations and the pilot sample size is $N = 2000$. ∎

Table 7.11: Cross-entropy method for basket call option

Strike price	$K = 50$		$K = 60$		$K = 70$	
	CE	plain	CE	plain	CE	plain
Estimate	1.2945	1.3240	0.0081	0.0090	3.8384×10^{-6}	0
S.E.	0.0071	0.0213	0.0001	0.0015	5.9022×10^{-8}	0
R.E.	0.55%	1.61%	1.11%	16.28%	1.54%	NaN

7.2.3 Initialization in Rare Event Simulation

We have seen from the previous discussion that the initialization in the iterative cross-entropy schemes becomes less straightforward when the simulation involves events with very small probabilities. The usual choice of $\hat{\theta}^0 = \theta^0$, where θ^0 corresponds to the original distribution, can be problematic in this context. For example, in the estimation of a very small probability $\mathbb{P}(X \in A)$, most likely the indicator $h(X_i) = 1_{\{X_i \in A\}}$ will be zero for all pilot samples X_i, rendering the basic cross-entropy scheme (7.7) or the first iteration of the updating rule (7.12) meaningless. However, it has been demonstrated that one can often resolve this issue by judiciously choosing a good initial tilting parameter $\hat{\theta}^0$ so that the general iterative scheme will converge to a good final tilting parameter within a few iterations.

In this section, we discuss a different, systematic initialization technique that takes the guesswork out of $\hat{\theta}^0$. It is indeed a *separate* iterative scheme to obtain a good initial tilting parameter $\hat{\theta}^0$. To fix ideas, consider the problem of estimating

$$\mu = E\left[H(X; \alpha) \cdot 1_{\{F(X) \geq \alpha\}} \right]$$

for some functions H and F and some constant α. We assume that H is nonnegative, and is strictly positive with high probability given $F(X) \geq \alpha$. The parameter α is called the *rarity parameter* — the bigger α is, the smaller the probability of $\{F(X) \geq \alpha\}$ becomes. Most of the estimation problems in financial engineering are of this type. For example, to estimate the price of an option, usually one can let H be the discounted payoff of the option, F some function of the underlying asset price, and α a parameter that represents the source of rarity (e.g., a large strike price or a low barrier price). Another example is that in the problem of estimating loss probabilities, often one can let H be one, F the size of the loss, and α a given large loss threshold. For notational simplicity, in all the subsequent discussions we assume that X is a random variable or a random vector with density f. The extension to discrete distributions is straightforward. The family of alternative sampling densities is denoted by $\{f_\theta(x)\}$ and $f(x) = f_{\theta^0}(x)$ for some θ^0.

The main idea is as follows. Recall that equation (7.11) with $h(x)$ replaced by $H(x; \alpha)1_{\{F(x) \geq \alpha\}}$ yields an optimal tilting parameter for estimating μ, regardless of the value of the tilting parameter ν. Now consider a different, iterative approach. Let

$$\alpha_1 \leq \alpha_2 \cdots \leq \alpha_m = \alpha$$

be an increasing sequence. Then solving equation (7.11), with $h(x)$ replaced by $H(x; \alpha_1)1_{\{F(x) \geq \alpha_1\}}$ and $\nu = \theta^0$, yields an optimal tilting parameter, say θ^1, for estimating

$$E\left[H(X; \alpha_1) \cdot 1_{\{F(X) \geq \alpha_1\}}\right].$$

With this new tilting parameter θ^1, solving equation (7.11) once more, with $h(x)$ replaced by $H(x; \alpha_2)1_{\{F(x) \geq \alpha_2\}}$ and $\nu = \theta^1$, yields an optimal tilting parameter for estimating

$$E\left[H(X; \alpha_2) \cdot 1_{\{F(X) \geq \alpha_2\}}\right].$$

Repeating this process, we will obtain an optimal tilting parameter for estimating μ in the end.

The motivation is that the cross-entropy method replaces equations like (7.11) by their stochastic versions such as (7.12). If α_{j+1} is chosen so that $F(Y) \geq \alpha_{j+1}$ happens with some moderate probability, given that Y has density f_{θ^j}, then the updating rule (7.12) will lead to a good estimate for θ^{j+1}. In other words, to reach the level α, a collection of intermediate levels are introduced, and the tilting parameter is adjusted gradually to make the transition from one level to the next no longer a rare event.

From these considerations, a natural choice of the sequence $\{\alpha_1, \ldots, \alpha_m\}$ is to recursively define α_{j+1} as the $(1 - \rho)$-quantile of the distribution of $F(Y)$, where Y has density f_{θ^j} and $0 < \rho < 1$ is some fraction. That is,

$$\alpha_{j+1} = \min\{x : \mathbb{P}(F(Y) \leq x) \geq 1 - \rho\}.$$

Even though these quantiles are nearly impossible to quantify analytically, they can be easily estimated from the samples of $F(Y)$. Moreover, since samples of $F(Y)$ will be drawn anyways to update the tilting parameter, these *same samples* can be used to estimate the quantile α_{j+1}.

To be more precise, denote by $\bar{\theta}^j$ the tilting parameter at the end of the j-th iteration. Let $\bar{\theta}^0 = \theta^0$. In the $(j+1)$-th iteration, one generates iid samples $Y_1, \ldots, Y_N$ from $f_{\bar{\theta}^j}$. Let $V_k = F(Y_k)$ and consider the order statistics

$$V_{(1)} \leq \cdots \leq V_{(N)}.$$

An estimate for the $(1-\rho)$-quantile is simply $\bar{\alpha}_{j+1} = V_{(N_0)}$, where $N_0 = [N(1-\rho)]$, the integer part of $N(1-\rho)$. Using these same samples, one can solve for $\bar{\theta}^{j+1}$ from equation (7.12) with $\nu = \bar{\theta}^j$ and $h(x)$ replaced by $H(x; \bar{\alpha}_{j+1})1_{\{F(x) \geq \bar{\alpha}_{j+1}\}}$. That is, $\bar{\theta}^{j+1}$ is the solution to the equation

$$0 = \frac{1}{N} \sum_{k=1}^{N} H(Y_k; \bar{\alpha}_{j+1})1_{\{F(Y_k) \geq \bar{\alpha}_{j+1}\}} \ell_{\bar{\theta}^j}(Y_k) \frac{\partial}{\partial \theta} \log f_{\theta}(Y_k). \qquad (7.17)$$

The iteration will end if $\bar{\alpha}_{j+1} \geq \alpha$. Note that in equation (7.17), only those samples Y_k that satisfy $F(Y_k) \geq \bar{\alpha}_{j+1}$ contribute to the updating of the tilting parameter. For this reason, these samples are called *elite samples*. By construction, the fraction of elite samples is approximately ρ. As before, equation (7.17) admits explicit solutions. In particular, in the case of normal distributions we have, analogous to Lemma 7.2,

$$\bar{\theta}^{j+1} = \frac{\sum_{k=1}^{N} H(Y_k; \bar{\alpha}_{j+1})1_{\{F(Y_k) \geq \bar{\alpha}_{j+1}\}} e^{-\langle \bar{\theta}^j, Y_k \rangle} Y_k}{\sum_{k=1}^{N} H(Y_k; \bar{\alpha}_{j+1})1_{\{F(Y_k) \geq \bar{\alpha}_{j+1}\}} e^{-\langle \bar{\theta}^j, Y_k \rangle}}. \qquad (7.18)$$

Below is the pseudocode. We would like to repeat that the final tilting parameter from this iterative scheme will *not* be used for importance sampling, but rather as the *initial tilting parameter* $\hat{\theta}^0$ for the general iterative cross-entropy algorithm. The parameter ρ should not be too small because it determines the fraction of samples that will be used for updating the tilting parameter. Nor should it be too large because otherwise the growth of $\bar{\alpha}_j$ will be too slow. In practice, ρ is usually chosen between 5% and 10%.

Pseudocode for the initialization of the iterative cross-entropy scheme:

> choose ρ between 5% and 10% and set $N_0 = [N(1-\rho)]$
> set $\bar{\theta}^0 = \theta^0$ and the iteration counter $j = 0$
> (*) generate N iid samples $Y_1, \ldots, Y_N$ from density $f_{\bar{\theta}^j}(x)$
> set $V_k = F(Y_k)$ and the order statistics $V_{(1)} \leq \cdots \leq V_{(N)}$
> set $\bar{\alpha}_{j+1} = V_{(N_0)}$ and $\bar{\theta}^{j+1}$ as the solution to (7.17)

set the iteration counter $j = j + 1$

if $\bar{\alpha}_j \geq \alpha$ set $\hat{\theta}^0 = \bar{\theta}^j$ and stop, otherwise go to step (*).

Example 7.9. Let us revisit Example 7.6. The call option price can be written as $v = E[H(X; \alpha)1_{\{F(X) \geq \alpha\}}]$, where X is a standard normal random variable and

$$\alpha = K, \quad F(X) = S_0 \exp\left\{\left(r - \frac{1}{2}\sigma^2\right)T + \sigma\sqrt{T}X\right\},$$

$$H(X; \alpha) = e^{-rT}\left(F(X) - \alpha\right)^+.$$

In the numerical experiment, we set $\rho = 10\%$ and $N = 2000$. Consequently, $N_0 = [2000 \times (1 - 0.10)] = 1800$. The other parameters are given by

$$S_0 = 50, \quad r = 0.05, \quad \sigma = 0.2, \quad T = 1.$$

The solution to equation (7.17) is given by (7.18), which is used to update $\bar{\theta}^j$. The initial tilting parameter is simply $\bar{\theta}^0 = \theta^0 = 0$. The number of iterations, the final rarity parameter $\bar{\alpha}$, and the final tilting parameter $\bar{\theta}$ are reported in Table 7.12.

Table 7.12: Initialization of the cross-entropy method for call option

	$K = 50$	$K = 60$	$K = 80$	$K = 100$	$K = 120$
# of iterations	1	1	2	2	3
Final $\bar{\alpha}$	66.3546	66.9077	100.8898	101.8036	150.7960
Final $\bar{\theta}$	2.1133	2.1284	3.8432	3.8940	5.7056

Note that the final tilting parameter $\bar{\theta}$ is close to be optimal for estimating the price of a call option with strike price $\bar{\alpha}$. From this point of view, the results are quite consistent with Table 7.9. One can use the general iterative cross-entropy scheme, with the initial tilting parameter $\hat{\theta}^0$ set as the final tilting parameter $\bar{\theta}$, to produce a nearly optimal tilting parameter for importance sampling within a couple of iterations. ■

7.3 Applications to Risk Analysis

Popular risk measures such as value-at-risk or expected tail loss involve tail probabilities that are usually very small. Importance sampling is particularly powerful in the estimation of such quantities. To fix ideas, let L

represent the loss and x be a large threshold. Consider the tail probability $\mathbb{P}(L > x)$ and the expected tail loss $E[L|L > x]$. A generic importance sampling estimate for the tail probability takes the form

$$\frac{1}{n} \sum_{i=1}^{n} 1_{\{\bar{L}_i > x\}} w_i,$$

where $\{\bar{L}_1, \ldots, \bar{L}_n\}$ are iid loss samples from some alternative distribution and $\{w_1, \ldots, w_n\}$ are the corresponding likelihood ratios. As for the expected tail loss, observe that it is closely related to the tail probability as one can easily show (see Example 1.10) that

$$E[L|L > x] = \frac{E[L1_{\{L>x\}}]}{\mathbb{P}(L > x)}. \tag{7.19}$$

Since

$$\frac{1}{n} \sum_{i=1}^{n} \bar{L}_i 1_{\{\bar{L}_i > x\}} w_i$$

is an unbiased estimate for the numerator $E[L1_{\{L>x\}}]$, it is natural to estimate the expected tail loss by

$$\frac{\sum_{i=1}^{n} \bar{L}_i 1_{\{\bar{L}_i > x\}} w_i}{\sum_{i=1}^{n} 1_{\{\bar{L}_i > x\}} w_i}.$$

The following lemma is concerned with the ratio of two sample means, which will be useful for constructing confidence intervals for the expected tail loss. For the sake of completeness, a proof is provided. If the reader is only interested in using it for confidence intervals, the proof can be safely skipped.

Lemma 7.3. *Assume that* $(X_1, Y_1), \ldots, (X_n, Y_n)$ *are iid random vectors such that*

$$E[X_i] = \mu, \quad E[Y_i] = \nu \neq 0.$$

Define

$$r = \frac{\mu}{\nu}, \quad R_n = \frac{\sum_{i=1}^{n} X_i}{\sum_{i=1}^{n} Y_i}, \quad \sigma_n^2 = \frac{\sum_{i=1}^{n} (X_i - R_n Y_i)^2}{(\sum_{i=1}^{n} Y_i)^2}.$$

Then as $n \to \infty$, R_n *converges to* r *with probability one and the distribution of*

$$\frac{R_n - r}{\sigma_n}$$

converges to the standard normal distribution.

PROOF. Without loss of generality, we assume $v > 0$; otherwise one can simply replace Y_i by $-Y_i$ and X_i by $-X_i$. Define $\bar{X}_n = \sum_{i=1}^{n} X_i/n$ and $\bar{Y}_n = \sum_{i=1}^{n} Y_i/n$. By the strong law of large numbers, with probability one

$$\lim_{n\to\infty} R_n = \lim_{n\to\infty} \frac{\bar{X}_n}{\bar{Y}_n} = \frac{\mu}{v} = r.$$

Moreover, it follows from straightforward computation, the strong law of large numbers, and the convergence of R_n to r, that with probability one

$$
\begin{aligned}
\lim_{n\to\infty} (\sqrt{n}\sigma_n \bar{Y}_n)^2 &= \lim_{n\to\infty} \frac{1}{n} \left(\sum_{i=1}^{n} X_i^2 - 2R_n \sum_{i=1}^{n} X_i Y_i + R_n^2 \sum_{i=1}^{n} Y_i^2 \right) \\
&= E[X^2] - 2rE[XY] + r^2 E[Y^2] \\
&= E[(X - rY)^2] \\
&= \mathrm{Var}(X - rY),
\end{aligned}
$$

where (X, Y) has the same distribution as (X_i, Y_i). Observe that

$$\frac{R_n - r}{\sigma_n} = \frac{\bar{X}_n - r\bar{Y}_n}{\sigma_n \bar{Y}_n} = \frac{\sqrt{n}(\bar{X}_n - r\bar{Y}_n)}{\sqrt{\mathrm{Var}(X - rY)}} \cdot \frac{\sqrt{\mathrm{Var}(X - rY)}}{\sqrt{n}\sigma_n \bar{Y}_n}.$$

The distribution of the first term converges to the standard normal distribution by the central limit theorem, whereas the second term converges to one with probability one (here we have used the assumption $v > 0$). The lemma follows readily. ∎

In order to illustrate the main idea, we focus on a popular class of portfolio credit risk models that uses *normal copula* to model the dependence structure among defaults. The special case with independent defaults has been investigated in Example 7.3 and will prove useful to the analysis of general credit risk models.

To be more concrete, assume that the model under consideration has m obligors. Let X_k be the default indicator for the k-th obligor and c_k the loss resulting from the default of the k-th obligor. Assuming for clarity that $c_k = 1$ for every k, the total loss L is

$$L = \sum_{k=1}^{m} c_k X_k = \sum_{k=1}^{m} X_k.$$

The default probability of the k-th obligor is assumed to be p_k. In other words, X_k is a Bernoulli random variable with parameter p_k, i.e.,

$$\mathbb{P}(X_k = 1) = p_k, \quad \mathbb{P}(X_k = 0) = 1 - p_k.$$

To model the dependence structure of the defaults, auxiliary random variables $\{Y_1, \ldots, Y_m\}$ are introduced so that

$$X_k = 1_{\{Y_k > y_k\}},$$
$$Y_k = \rho_k Z + \sqrt{1 - \rho_k^2}\varepsilon_k,$$

where $y_k \in \mathbb{R}$ and $-1 < \rho_k < 1$ for each k, and $\{Z, \varepsilon_1, \ldots, \varepsilon_m\}$ are iid standard normal random variables. Since Y_k is a standard normal random variable itself and $\mathbb{P}(X_k = 1) = p_k$, it follows that

$$y_k = \Phi^{-1}(1 - p_k) = -\Phi^{-1}(p_k).$$

Observe that X_k's are *dependent* because of the single common factor Z. In general, it is possible to introduce multiple common factors. However, as far as importance sampling is concerned, the difference is only notational.

We should first discuss the estimation of the tail probability $\mathbb{P}(L > x)$. It follows from the tower property that $\mathbb{P}(L > x) = E[h(Z)]$, where

$$h(z) = \mathbb{P}(L > x | Z = z).$$

Therefore, we can divide the estimation into two subproblems: (1) estimate $h(z)$ for $z \in \mathbb{R}$; (2) estimate $E[h(Z)]$ where Z is a standard normal random variable.

The key observation in estimating $h(z)$ is that $\{Y_1, \ldots, Y_m\}$ are *independent* conditional on Z. The conditional independence implies that $h(z)$ equals the loss probability of a credit risk model with independent defaults, where the default probability of the k-the obligor is (abusing notation)

$$p_k(z) = \mathbb{P}(Y_k > y_k | Z = z) = \Phi\left(-\frac{y_k - \rho_k z}{\sqrt{1 - \rho_k^2}}\right). \tag{7.20}$$

Thus it follows from Example 7.3 that an efficient importance sampling scheme for estimating $h(z)$ is to use the alternative sampling distribution under which the default probability of the k-th obligor is given by

$$\bar{p}_k(z) = \frac{p_k(z)e^{\theta^*(z)}}{1 + p_k(z)(e^{\theta^*(z)} - 1)}, \tag{7.21}$$

where the tilting parameter $\theta^*(z) \in \mathbb{R}$ is defined in the following fashion. If

$$x > \sum_{k=1}^{m} p_k(z) = E\left[\sum_{k=1}^{m} X_k \,\middle|\, Z = z\right] = E[L | Z = z], \tag{7.22}$$

then $\theta^*(z)$ is defined to be the unique positive solution to an equation analogous to (7.3), namely,

$$x = \frac{d}{d\theta}\phi(\theta;z) = \sum_{k=1}^{m} \frac{p_k(z)}{p_k(z) + [1 - p_k(z)]e^{-\theta}}, \qquad (7.23)$$

where

$$\phi(\theta;z) = \sum_{k=1}^{m} \log[1 + p_k(z)(e^{\theta} - 1)].$$

When (7.22) fails, $h(z)$ is not a small quantity and it suffices to use the plain Monte Carlo scheme, or equivalently, define $\theta^*(z) = 0$.

Importance sampling can also be applied to the estimation of $E[h(Z)]$. It is particularly important to do so when the correlations among $\{Y_k\}$ are strong, in which case a large loss is more likely due to a large value of Z, and thus altering the sampling distribution of Z becomes more imperative. Assume as usual that the family of alternative sampling densities is $\{f_\mu(x) : \mu \in \mathbb{R}\}$, where f_μ is the density of $N(\mu, 1)$. The original distribution of Z corresponds to $\mu = \mu^0 = 0$. A good tilting parameter μ can be determined by the general iterative cross-entropy method. The only trouble here is that h is not explicitly known and has be to be replaced by an approximation or an estimate, say $\hat{h}$, which leads to the following updating rule suggested by Lemma 7.2:

$$\hat{\mu}^{j+1} = \frac{\sum_{k=1}^{N} \hat{h}(\bar{Z}_k)e^{-\hat{\mu}^j \bar{Z}_k} \bar{Z}_k}{\sum_{k=1}^{N} \hat{h}(\bar{Z}_k)e^{-\hat{\mu}^j \bar{Z}_k}},$$

where $\bar{Z}_1, \ldots, \bar{Z}_N$ are iid pilot samples from $N(\hat{\mu}^j, 1)$. The initial tilting parameter is simply set to be $\hat{\mu}^0 = \mu^0 = 0$.

It remains to determine $\hat{h}$. For a given z, it is possible to take $\hat{h}(z)$ to be the importance sampling estimate of $h(z)$ as described previously. The drawback is that it will require nontrivial computational resource unless the sample size for simulating $h(z)$ is very small. We will adopt another approach which does not involve simulation. Following exactly the same steps that led to the upper bound (7.2), one can establish an upper bound for $h(z)$, that is, for any $\theta \geq 0$

$$h(z) = P(L > x | Z = z) \leq e^{-\theta x + \phi(\theta;z)}.$$

In particular,

$$\hat{h}(z) = \exp\{-\theta^*(z)x + \phi(\theta^*(z);z)\}$$

is an upper bound, as well as a good approximation, of h; see Exercise 7.8. Below is the pseudocode.

Pseudocode for the iterative cross-entropy method for $\hat{\mu}$:

> initialize $\hat{\mu}^0 = 0$ and set the iteration counter $j = 0$
> (*) for $i = 1, 2, \ldots, N$
> > generate sample $\bar{Z}_i$ from $N(\hat{\mu}^j, 1)$
> > calculate $P_k = p_k(\bar{Z}_i)$ for $k = 1, \ldots, m$ from (7.20)
> > set $l = P_1 + \cdots + P_m$
> > if $l < x$ solve $\theta = \theta^*(\bar{Z}_i)$ by the bisection method from (7.23);
> > > otherwise set $\theta = \theta^*(\bar{Z}_i) = 0$
> > set $\hat{h}_i = \hat{h}(\bar{Z}_i) = \exp\{-\theta x + \phi(\theta; \bar{Z}_i)\}$
> > set $\hat{\mu}^{j+1} = \sum_{k=1}^N \hat{h}_k e^{-\hat{\mu}^j \bar{Z}_k} \bar{Z}_k / \sum_{k=1}^N \hat{h}_k e^{-\hat{\mu}^j \bar{Z}_k}$
> > set the iteration counter $j = j + 1$
> > if $j = \text{IT\_NUM}$ set $\hat{\mu} = \hat{\mu}^j$, otherwise go to step (*)

Once the tilting parameter $\hat{\mu}$ is obtained, it is easy to construct an importance sampling estimate for the tail probability $\mathbb{P}(L > x)$ using the sample average of iid copies of

$$H = 1_{\{L > x\}} \cdot \exp\left\{-\hat{\mu}\bar{Z} + \frac{1}{2}\hat{\mu}^2\right\} \cdot \prod_{k=1}^m \left(\frac{p_k(\bar{Z})}{\bar{p}_k(\bar{Z})}\right)^{\bar{X}_k} \left(\frac{1 - p_k(\bar{Z})}{1 - \bar{p}_k(\bar{Z})}\right)^{1 - \bar{X}_k},$$

where $\bar{Z}$ has distribution $N(\hat{\mu}, 1)$, $p_k(\bar{Z})$ is given by (7.20), $\bar{p}_k(\bar{Z})$ is given by (7.21), $\bar{X}_k$'s are independent Bernoulli random variables conditional on $\bar{Z}$ with

$$\mathbb{P}(\bar{X}_k = 1 | \bar{Z}) = \bar{p}_k(\bar{Z}), \quad k = 1, \ldots, m,$$

and

$$\bar{L} = \sum_{k=1}^m \bar{X}_k.$$

In this algorithm, importance sampling is used for estimating both $h(z)$ and $E[h(Z)]$.

The above scheme can be modified slightly to yield an estimate for the expected tail loss. As we have mentioned previously, the sample average of iid copies of

$$\bar{L} \cdot H$$

is an unbiased estimate for $E[L 1_{\{L > x\}}]$. Therefore, a natural estimate for $E[L | L > x]$ is simply

$$R_n = \frac{\sum_{i=1}^n \bar{L}_i H_i}{\sum_{i=1}^n H_i},$$

where H_i's are iid copies of H and $\bar{L}_i$'s are the corresponding losses. The standard error associated with this estimate, by Lemma 7.3, is

$$\text{S.E.} = \sqrt{\frac{\sum_{i=1}^{n}(\bar{L}_i H_i - R_n H_i)^2}{(\sum_{i=1}^{n} H_i)^2}}.$$

In other words, a $(1 - \alpha)$ confidence interval of the expected tail loss is approximately

$$R_n \pm z_{\alpha/2} \cdot \text{S.E.}$$

where $z_{\alpha/2}$ is determined by $\Phi(-z_{\alpha/2}) = \alpha/2$. Below is the pseudocode for estimating the tail probability and the expected tail loss.

Pseudocode for the tail probability and expected tail loss:

use the iterative cross-entropy method to obtain a tilting parameter $\hat{\mu}$

for $i = 1, 2, \ldots, n$

    generate $\bar{Z}$ from $N(\hat{\mu}, 1)$

    calculate $P_k = p_k(\bar{Z})$ for $k = 1, \ldots, m$ from (7.20)

    set $l = P_1 + \cdots + P_m$

    if $l < x$ solve $\theta = \theta^*(\bar{Z})$ by the bisection method from (7.23);

        otherwise set $\theta = \theta^*(\bar{Z}) = 0$

    calculate $\bar{P}_k = \bar{p}_k(\bar{Z})$ for $k = 1, \ldots, m$ from (7.21)

    generate $\bar{X}_k$ from Bernoulli with parameter $\bar{P}_k$ for $k = 1, \ldots, m$

    set $\bar{L}_i = \bar{X}_1 + \cdots + \bar{X}_m$

    if $\bar{L}_i > x$ set $H_i = e^{-\mu \bar{Z} + \mu^2/2} \cdot \prod_{k=1}^{m} \left(\frac{P_k}{\bar{P}_k}\right)^{\bar{X}_k} \left(\frac{1 - P_k}{1 - \bar{P}_k}\right)^{1 - \bar{X}_k};$

    otherwise set $H_i = 0$

compute the estimate for tail probability $\hat{v} = \dfrac{1}{n} \sum_{i=1}^{n} H_i$

compute the standard error of $\hat{v} = \sqrt{\dfrac{1}{n(n-1)} \left(\sum_{i=1}^{n} H_i^2 - n\hat{v}^2\right)}$

compute the estimate for expected tail loss $\hat{r} = \dfrac{1}{n\hat{v}} \sum_{i=1}^{n} \bar{L}_i H_i$

compute the standard error of $\hat{r} = \dfrac{1}{n\hat{v}} \sqrt{\sum_{i=1}^{n} (\bar{L}_i H_i - \hat{r} H_i)^2}.$

If we fix the tilting parameter $\hat{\mu} = 0$ in the above importance sampling scheme, then it is equivalent to using importance sampling for the estimation of $h(z)$ but using plain Monte Carlo for the estimation of $E[h(Z)]$. We call such a scheme "partial importance sampling." We should include it in our numerical experiment to investigate the effect of the correlations among defaults on its performance. We expect that as the correlations grow stronger, the partial importance sampling will become less effective because a large loss is more likely due to a large value of the common factor Z.

Numerical Experiment. Consider a credit risk model where the parameters are given by

$$m = 1000, \quad p_k = 0.01 \cdot [1 + e^{-k/m}], \quad \rho_k = \rho, \quad n = 10000.$$

For the iterative cross-entropy method, we use IT_NUM = 5 iterations and $N = 2000$ pilot samples for each iteration (in the actual simulation, the convergence is attained within one iteration, which means that the basic cross-entropy scheme would have sufficed). The parameter ρ captures the strength of the correlations among defaults. A larger value of $|\rho|$ means stronger correlations. When $\rho = 0$, the defaults are independent.

Table 7.13: Credit risk model: IS versus Plain MC versus Partial IS

$x = 26, \ \rho = 0.05$	TProb	R.E.	ETL $- x$	R.E.	$\hat{\mu}$
IS	0.0180	1.59%	2.6180	1.06%	0.9785
Plain MC	0.0196	7.07%	2.7296	6.04%	0
Partial IS	0.0181	3.00%	2.5541	1.99%	0
$x = 30, \ \rho = 0.1$	TProb	R.E.	ETL $- x$	R.E.	$\hat{\mu}$
IS	0.0159	1.53%	3.7131	1.30%	1.5803
Plain MC	0.0170	7.60%	3.3412	5.78%	0
Partial IS	0.0168	5.17%	3.5487	4.43%	0
$x = 60, \ \rho = 0.3$	TProb	R.E.	ETL $- x$	R.E.	$\hat{\mu}$
IS	0.0156	1.49%	16.2838	1.25%	2.2538
Plain MC	0.0158	7.89%	15.9114	6.98%	0
Partial IS	0.0140	7.86%	16.2384	7.79%	0
$x = 145, \ \rho = 0.6$	TProb	R.E.	ETL $- x$	R.E.	$\hat{\mu}$
IS	0.0157	1.52%	76.4701	1.31%	2.4254
Plain MC	0.0161	7.82%	69.7888	7.79%	0
Partial IS	0.0162	7.68%	85.2236	8.14%	0

Table 7.14: Credit risk model: IS versus Plain MC versus Partial IS

$x = 40$, $\rho = 0.1$	TProb	R.E.	ETL $- x$	R.E.	$\hat{\mu}$
IS	5.8800×10^{-4}	1.76%	3.3147	1.30%	2.5216
Plain MC	6.0000×10^{-4}	40.81%	2.8333	17.48%	0
Partial IS	7.8932×10^{-4}	21.09%	4.3622	19.49%	0
$x = 50$, $\rho = 0.1$	TProb	R.E.	ETL $- x$	R.E.	$\hat{\mu}$
IS	1.4478×10^{-5}	1.96%	3.0765	1.33%	3.3562
Plain MC	0	NaN	NaN	NaN	0
Partial IS	8.3848×10^{-6}	52.04%	2.7119	37.39%	0

The numerical results are reported in Tables 7.13 and 7.14. The entry "TProb" means the estimate for the tail probability $\mathbb{P}(L > x)$, while "ETL $-$ x" means the estimate for $E[L|L > x] - x$.

In Table 7.13, we push up the threshold x as ρ increases in order for the tail probability to remain roughly the same magnitude. The efficiency of partial importance sampling clearly deteriorates as the correlations grow stronger. Note that the performance of partial importance sampling is indistinguishable from that of the plain Monte Carlo scheme when ρ is large, which suggests that using importance sampling for estimating $h(z)$ alone does not achieve much variance reduction overall. This justifies our intuition that a large loss is most likely due to a large value of the common factor Z. In Table 7.14, we fix ρ and let x vary. It is clear that only the importance sampling scheme consistently yields accurate estimates even when the tail probability becomes really small. ∎

Exercises

Pen-and-Paper Problems

7.1 Write down the importance sampling estimates for the following quantities:

(a) $\mathbb{P}(X \geq m\alpha)$, where X is a binomial random variable with parameters (m, p) and $\alpha \in (0, 1)$ is a given constant. The alternative sampling distribution is binomial with parameters $(m, \bar{p})$.

(b) $E[h(X_1, \ldots, X_m)]$, where $(X_1, \ldots, X_m)$ is an m-dimensional standard normal random vector. The alternative sampling distribution is assumed to be $N(\theta, I_m)$ for some $\theta = (\theta_1, \ldots, \theta_m) \in \mathbb{R}^m$.

(c) $E[h(X_1, \ldots, X_m)]$, where $(X_1, \ldots, X_m)$ is a jointly normal random vector with mean μ and nonsingular covariance matrix Σ. *Hint:* Use (b) and Cholesky's factorization.

(d) $\mathbb{P}(X_1 + \cdots + X_m \geq a)$, where $(X_1, \ldots, X_m)$ is a random vector with joint density $f(x_1, \ldots, x_m)$. The alternative sampling density is assumed to be $g(x_1, \ldots, x_m)$.

7.2 Let X be a standard normal random variable. Assume that we wish to use importance sampling to estimate, for a given positive constant a,

$$E\left(e^{a\sqrt{X}} 1_{\{X \geq 0\}}\right).$$

The alternative sampling distribution is $N(\theta, 1)$ for some $\theta \in \mathbb{R}$. Use the mode matching method to determine the best θ.

7.3 The purpose of this exercise is to justify the method of mode matching for normal distributions. Suppose that one is interested in estimating the tail probability

$$\mathbb{P}(Z \geq b)$$

for a standard normal random variable Z and a large threshold b.

(a) Write down the importance sampling estimate with the alternative sampling distribution $N(\theta, 1)$.

(b) Compute the variance of this estimate.

(c) Assume that the variance is minimized at $\theta = \theta^*$. Write down the equation that θ^* satisfies.

(d) Solve the equation to obtain θ^*, using the extremely accurate approximation [4] that for large x

$$\Phi(-x) \approx \frac{\phi(x)}{x}, \quad \phi(x) = \frac{1}{\sqrt{2\pi}} e^{-x^2/2}.$$

7.4 A Cautionary Example. The purpose of this exercise is to show that the method of mode matching will not work universally. Indeed, it can be much worse than the plain Monte Carlo. The setup is the same as Exercise 7.3 except that the quantity of interest is

$$\mathbb{P}(Z \geq b \text{ or } Z \leq -2b)$$

The alternative sampling distribution is assumed to be $N(\theta, 1)$.

(a) Use the method of mode matching to determine θ.

(b) Write down the importance sampling estimate H.

(c) Compute $E[H^2]$ explicitly and use the same approximation in Exercise 7.3 to argue that

$$\lim_{b \to \infty} E[H^2] = \infty.$$

The method of mode matching should be used with caution. The same can be said to the cross-entropy method. Indeed, when the problem exhibits certain kind of nonconvexity, looking for an alternative sampling distribution within the exponential tilt family may not be sufficient [9]. This counterexample was first constructed in [13]. See also Exercise 7.L.

7.5 Determine the exponential tilt family for the random variable X.

(a) X is normal with mean 0 and variance σ^2.

(b) X is Bernoulli with parameter p.

(c) X is Poisson with parameter λ.

(d) X is exponentially distributed with rate λ.

7.6 The exponential tilt family for a general random vector can be defined in the same fashion. Let $X = (X_1, \dots, X_m)$ be an m-dimensional random vector.

(a) *X is continuous:* Let f denote the density of X. The exponential tilt family consists of density functions of the form

$$f_\theta(x) = \frac{1}{E[e^{\langle \theta, X \rangle}]} e^{\langle \theta, x \rangle} f(x)$$

for some $\theta \in \mathbb{R}^m$.

(b) *X is discrete:* Let p denote the probability mass function of X, that is, $\mathbb{P}(X = x) = p(x)$. The exponential tilt family consists of probability mass functions of the form

$$p_\theta(x) = \frac{1}{E[e^{\langle \theta, X \rangle}]} e^{\langle \theta, x \rangle} p(x)$$

for some $\theta \in \mathbb{R}^m$.

Determine the exponential tilt family for X when (i) X is $N(0, I_m)$; (ii) the components $X_1, \ldots, X_m$ are independent random variables.

7.7 Suppose that $X = Y_1 + \cdots + Y_m$ where Y_k's are independent discrete random variables. Fix an arbitrary θ. Let $\bar{Y}_1, \ldots, \bar{Y}_m$ be independent with distribution

$$\mathbb{P}(\bar{Y}_k = y) = \frac{1}{E[e^{\theta Y_k}]} e^{\theta y} \mathbb{P}(Y_k = y)$$

for each k. That is, $\bar{Y}_k$ has the exponential tilt distribution of Y_k with parameter θ. Show that $\bar{X} = \bar{Y}_1 + \cdots + \bar{Y}_m$ has the exponential tilt distribution of X with parameter θ. That is,

$$\mathbb{P}(\bar{X} = x) = \frac{1}{E[e^{\theta X}]} e^{\theta x} \mathbb{P}(X = x).$$

In other words, if one wishes to use the exponential tilt distribution of X with parameter θ for importance sampling, it suffices to use the exponential tilt distribution on each component Y_k with the *same* θ. This result justifies the alternative sampling distribution used in Example 7.3. Not surprisingly, a similar result holds when Y_k's are continuous random variables.

7.8 Consider the credit risk model in Example 7.3. Show that equation (7.3) is equivalent to

$$x = E[\bar{L}] = \sum_{k=1}^{m} \bar{p}_k.$$

Show that in this case, the standard deviation of $\bar{L}$ is bounded from above by $\sqrt{x}$, which is much smaller compared to x when x is large. Use this to explain why (7.2) gives a good upper bound for $E[H^2]$ when θ is taken to be the solution to equation (7.3).

7.9 Consider a slight generalization of the credit risk model in Example 7.3, where c_k's are not always one. Show that $\bar{L}$ has the exponential tilt distribution of L with parameter θ, where

$$\bar{L} = \sum_{k=1}^{m} c_k Y_k$$

and Y_k's are independent Bernoulli random variables with parameter

$$\bar{p}_k = \mathbb{P}(Y_k = 1) = \frac{1}{E[e^{\theta c_k X_k}]} e^{\theta c_k} p_k = \frac{p_k e^{\theta c_k}}{1 + p_k (e^{\theta c_k} - 1)}.$$

Hint: Apply Exercise 7.7 with $Y_k = c_k X_k$.

7.10 Let $f(x)$ and $g(x)$ be two density functions. Define the Kullback–Leibler cross entropy or relative entropy by

$$R(g \| f) = \int_{\mathbb{R}} \log \frac{g(x)}{f(x)} \cdot g(x) \, dx.$$

Show that $R(g\|f) \geq 0$ and $R(g\|f) = 0$ if and only if $g(x) = f(x)$. Similarly, if $p(x_i)$ and $q(x_i)$ are two probability mass functions, then the relative entropy is defined as

$$R(q\|p) = \sum_i \log \frac{q(x_i)}{p(x_i)} \cdot q(x_i)$$

Again, show that $R(q\|p) \geq 0$ and $R(q\|p) = 0$ if and only if $q = p$. *Hint:* Let X be a random variable with density f. Define $Y = g(X)/f(X)$. Argue that the function $h(x) = x \log x$ is convex and apply the Jensen's inequality (Exercise 1.23) on $R(g\|f) = E[h(Y)]$. The discrete case can be proved similarly.

7.11 Assume that X is a random variable with density f, and $\{f_\theta\}$ is the exponential tilt family defined in Section 7.1.3. That is,

$$f_\theta(x) = \frac{1}{E[e^{\theta X}]} e^{\theta x} f(x).$$

Let $H(\theta) = \log E[e^{\theta X}]$. Show that the solution $\hat{\theta}$ to equation (7.7) is determined by

$$H'(\hat{\theta}) = \frac{\sum_{k=1}^N h(X_k) X_k}{\sum_{k=1}^N h(X_k)}.$$

Similarly, the solution $\hat{\theta}^{j+1}$ to equation (7.12) is determined by

$$H'(\hat{\theta}^{j+1}) = \frac{\sum_{k=1}^N h(Y_k) \ell_{\hat{\theta}^j}(Y_k) Y_k}{\sum_{k=1}^N h(Y_k) \ell_{\hat{\theta}^j}(Y_k)}.$$

Argue that the exactly same formulas hold when X is a discrete random variable.

7.12 Consider the problem of estimating $E[h(X)]$, where X is a binomial random variable with parameters (m, p) and h is nonnegative. Assuming that the alternative sampling distribution is binomial with parameters (m, θ), write down the solution to the basic cross-entropy scheme (7.7) and to the general iterative scheme (7.12).

7.13 Consider the credit risk model in Example 7.3. Assuming that the iterative cross-entropy scheme is to be used in order to obtain a good tilting parameter θ, show that the updating rule, or the solution to equation (7.12), is determined by

$$\phi'(\hat{\theta}^{j+1}) = \frac{\sum_{i=1}^N \bar{L}_i e^{-\hat{\theta}^j \bar{L}_i} 1_{\{L_i > x\}}}{\sum_{i=1}^N e^{-\hat{\theta}^j \bar{L}_i} 1_{\{L_i > x\}}},$$

where $\bar{L}_1, \ldots, \bar{L}_N$ are iid copies of $\bar{L} = Y_1 + \cdots + Y_m$ and Y_k's are independent Bernoulli random variables with parameter

$$\bar{p}_k = \mathbb{P}(Y_k = 1) = \frac{p_k e^{\hat{\theta}_j}}{1 + p_k(e^{\hat{\theta}_j} - 1)}.$$

7.14 Let $Z = (Z_1, \ldots, Z_m)$ be an m-dimensional standard normal random vector. Denote the exponential tilt family of Z by $\{f_\theta(x) : \theta \in \mathbb{R}^m\}$, where f_θ is the density function of $N(\theta, I_m)$ for $\theta = (\theta_1, \ldots, \theta_m) \in \mathbb{R}^m$. In some applications, one can restrict the alternative sampling distribution to a subset of the exponential tilt family where θ takes the form $\theta = (\mu, \ldots, \mu)$ for some $\mu \in \mathbb{R}$. Under this assumption, write down the solution to the basic cross-entropy scheme (7.7) and to the general iterative scheme (7.12).

MATLAB® Problems

7.A Compare importance sampling with the plain Monte Carlo scheme in the estimation of $\mathbb{P}(X \geq b)$.

(a) X is binomial with parameters (m, p) and $b = m\alpha$ with $p < \alpha < 1$. The alternative sampling distribution is binomial with parameters (m, α). Report your results for $p = 0.5, \alpha = 0.9$, and $m = 10, 100, 1000$, respectively. Sample size is $n = 10000$.

(b) X is exponential with rate λ and $b = a/\lambda$ with $a > 1$. The alternative sampling distribution is exponential with rate $1/b$. Report your results for $\lambda = 1$ and $a = 2, 5, 10$, respectively. Sample size is $n = 10000$.

7.B Let X, Y, Z be iid standard normal random variables. Use the basic cross-entropy scheme to estimate

$$\mathbb{P}(\min\{X + Y, Y + 2Z + 1\} \geq b)$$

The sample size is $n = 10000$. The pilot sample size is $N = 2000$.

(a) Compare with plain Monte Carlo for $b = 1, 2, 3$, respectively.

(b) Let $b = 5$. Does the basic cross-entropy method still work? Why?

In Exercises 7.C — 7.G, the underlying asset price is assumed to be a geometric Brownian motion under the risk-neutral probability measure:

$$S_t = S_0 \exp\left\{\left(r - \frac{1}{2}\sigma^2\right)t + \sigma W_t\right\},$$

where W is a standard Brownian motion and r is the risk-free interest rate.

7.C Consider the binary option in Example 7.1. Write a function to estimate the option price by the iterative cross-entropy method. The function should have input parameters S_0, r, σ, T, K, sample size n, pilot sample size N, and the number of iterations IT_NUM. Explain your choice of the initial tilting parameter. Report your estimate, relative error, and the final tilting parameter for

$$S_0 = 50, \ r = 0.01, \ \sigma = 0.1, \ T = 1, \ n = 10000, \ N = 2000, \ \text{IT\_NUM} = 5,$$

and $K = 50, 60, 70, 80$, respectively. How many iterations are actually needed for the tilting parameters to converge?

7.D Write a function to estimate the price of a put option with maturity T and strike price K, via importance sampling. Use the mode matching method to find a good tilting parameter. The function should have input parameters S_0, r, σ, K, T, n. Report your estimate, standard error, and the tilting parameter for

$$r = 0.1, \ \sigma = 0.2, \ T = 1, \ K = 30, \ n = 10000,$$

and $S_0 = 30, 50, 70$, respectively.

7.E Repeat Exercise 7.D, but use the iterative cross-entropy method to find a good tilting parameter. The function should have two additional input parameters, namely, the pilot sample size N and the number of iterations IT_NUM. Explain your choice of the initial tilting parameter. Set $N = 2000$ and IT_NUM $= 5$ in your numerical simulation.

7.F Consider a lookback call option with fixed strike price K and maturity T, whose payoff is

$$\left(\max_{i=1,\dots,m} S_{t_i} - K \right)^+$$

for some given dates $0 < t_1 < \cdots < t_m = T$. The plain Monte Carlo scheme or the basic cross-entropy method fails when the strike price K is large. Write a function to estimate the option price by the iterative cross-entropy method. The function should have input parameters $S_0, r, \sigma, K, T, m, (t_1, \dots, t_m)$, sample size n, pilot sample size N, and the number of iterations IT_NUM. Explain your choice of the initial tilting parameter. Report your estimate, standard error, and the final tilting parameter for

$$S_0 = 50, \ r = 0.02, \ \sigma = 0.2, \ T = 0.5, \ m = 12, \ t_i = iT/m,$$

$$n = 10000, \ N = 2000, \ \text{IT\_NUM} = 5,$$

and $K = 60, 80, 100$, respectively.

7.G Consider a binary down-and-in barrier option with maturity T and barrier b, whose payoff is

$$1_{\{\min(S_{t_1}, \dots, S_{t_m}) \le b\}}$$

for a given set of dates $0 < t_1 < \cdots < t_m = T$. The plain Monte Carlo or the basic cross-entropy scheme fails when the barrier b is much lower than S_0. Note that the payoff is a function of $Z = (Z_1, \ldots, Z_m)$, where

$$Z_k = (W_{t_k} - W_{t_{k-1}})/\sqrt{t_k - t_{k-1}}, \quad 1 \le k \le m,$$

are iid standard normal random variables. Design an iterative cross-entropy scheme, with the alternative sampling distribution restricted to the subset of the exponential tilt family of Z given by Exercise 7.14. Explain your choice of the initial tilting parameter. Report your estimate, standard error, and the final tilting parameter for

$$S_0 = 50, \ r = 0.02, \ \sigma = 0.2, \ T = 1, \ m = 100, \ t_i = iT/m,$$

$$n = 10000, \ N = 2000, \ \text{IT\_NUM} = 5,$$

and $b = 40, 30, 20$, respectively.

7.H The setup is similar to Exercise 6.F. Let X and Y be the prices of two underlying stocks. Assume that they are both geometric Brownian motions under the risk-neutral probability measure and

$$X_T = X_0 \exp\left\{ \left(r - \frac{1}{2}\sigma_1^2 \right) T + \sigma_1 \sqrt{T} Z_1 \right\},$$

$$Y_T = Y_0 \exp\left\{ \left(r - \frac{1}{2}\sigma_2^2 \right) T + \sigma_2 \sqrt{T} Z_2 \right\},$$

where Z_1 and Z_2 are two independent standard normal random variables. We wish to estimate the price of a basket call option with maturity T and payoff

$$h(Z_1, Z_2) = (c_1 X_T + c_2 Y_T - K)^+$$

by importance sampling. The alternative sampling distribution is assumed to be $N(\theta, I_2)$ for some $\theta = (\theta_1, \theta_2) \in \mathbb{R}^2$.

(a) The mode matching method sets the tilting parameter (θ_1, θ_2) as the maximizer of $h(z_1, z_2)f(z_1, z_2)$, where f is the joint density function of $Z = (Z_1, Z_2)$. It is equivalent to maximizing

$$h(z_1, z_2) \exp\left\{ -\frac{1}{2}(z_1^2 + z_2^2) \right\}.$$

Find the maximizer (z_1^*, z_2^*) by the bisection method. *Hint:* Taking partial derivatives over z_1 and z_2 leads to two equations with two unknowns. Reduce the system of equations to a single equation with one unknown.

(b) Write a MATLAB function to estimate the call option price by importance sampling. The function should have input parameters r, σ_1, σ_2, X_0, Y_0, T, K, c_1, c_2, and sample size n. Report your estimate, standard error, and the tilting parameter (θ_1, θ_2) for

$$r = 0.1, \ \sigma_1 = 0.2, \ \sigma_2 = 0.3, \ X_0 = Y_0 = 50, \ T = 1, \ c_1 = c_2 = 0.5,$$

and strike $K = 80, 100, 120$, respectively. Sample size is $n = 10000$.

7.I Consider a multi-asset basket put option with maturity T and strike price K. The option payoff is

$$\left(K - \sum_{i=1}^{d} c_i S_T^{(i)} \right)^+.$$

Assume that under the risk-neutral probability measure, the underlying stock prices are modeled by geometric Brownian motions:

$$S_t^{(i)} = S_0^{(i)} \left\{ \left(r - \frac{1}{2}\sigma_i^2 \right) t + \sigma_i W_t^{(i)} \right\}, \quad i = 1, \dots, d,$$

where $W = (W^{(1)}, \dots, W^{(d)})$ is a d-dimensional Brownian motion with covariance matrix $\Sigma = [\Sigma_{ij}]$ such that $\Sigma_{ii} = 1$ for all i. Write a function to estimate the option price by the iterative cross-entropy method. The function should have input parameters $d, r, K, T, \Sigma, S_0^{(i)}, \sigma_i, c_i$, sample size n, pilot sample size N, and the number of iterations IT_NUM. Explain your choice of the initial tilting parameter. Report your estimate, standard error, and the final tilting parameter for

$$d = 3, \ S_0^{(1)} = 50, \ S_0^{(2)} = 45, \ S_0^{(3)} = 55, \ r = 0.03, \ T = 1,$$

$$\sigma_1 = 0.2, \ \sigma_2 = \sigma_3 = 0.3, \ \Sigma = \begin{bmatrix} 1 & 0.2 & -0.2 \\ 0.2 & 1 & 0.5 \\ -0.2 & 0.5 & 1 \end{bmatrix},$$

$$c_1 = 0.4, \ c_2 = c_3 = 0.3, \ n = 10000, \ N = 2000, \ \text{IT\_NUM} = 5,$$

and $K = 50, 40, 30$, respectively.

7.J Consider a slight generalization of the simple credit risk model in Example 7.3, where the losses c_k are not always one. Write a function to estimate the tail probability $\mathbb{P}(L > x)$ and the expected tail loss $E[L|L > x]$ by importance sampling. The alternative sampling distribution is given by Exercise 7.9. Use the mode matching method to determine a good tilting parameter θ. The function should have input parameters m, $(p_1, \dots, p_m)$, $(c_1, \dots, c_m)$, x, and sample size n. Report your estimate, standard error, and the tilting parameter for

$$m = 1000, \ p_k = 0.01 \cdot [1 + e^{-k/m}], \ c_k = 1 + e^{k/m}, \ n = 10000,$$

and $x = 50, 70, 90$, respectively.

7.K Consider the credit risk model in Example 7.3. Use the initialization technique described in Section 7.2.3 to find an initial tilting parameter for the cross-entropy scheme. Set $\rho = 10\%$, $\bar{\theta}^0 = 0$, and pilot sample size $N = 2000$. Report the final tilting parameter $\bar{\theta}$ and the number of iterations performed. Let $\hat{\theta}^0 = \bar{\theta}$ be the initial tilting parameter for the iterative cross-entropy scheme, and report your final tilting parameter $\hat{\theta}$ for importance sampling, with pilot sample size $N = 2000$ and the number of iterations IT_NUM $= 5$. *Hint:* Apply the results in Exercise 7.13.

7.L This exercise is to reinforce numerically what Exercise 7.4 has demonstrated. Consider a very simple problem of estimating

$$\mathbb{P}(Z \in A),$$

where Z is a standard normal random variable and $A = (-\infty, a] \cup [b, \infty)$ is a nonconvex set. Compare the following algorithms:

i. The plain Monte Carlo.

ii. Importance sampling using the mode matching method.

iii. Importance sampling using the iterative cross-entropy method with the initial tilting parameter $\hat{\theta}^0 = a$.

iv. Importance sampling using the iterative cross-entropy method with the initial tilting parameter $\hat{\theta}^0 = b$.

Perform ten sets of simulations for each scheme given $a = -2.2$ and $b = 2$, with sample size $n = 10000$, pilot sample size $N = 2000$, and the number of iterations IT_NUM $= 5$. Report your estimates, standard errors, and 95% confidence intervals. Describe your findings. From the simulation, you will see clearly that the standard error of an importance sampling estimate is not always trustworthy when the problem involves nonconvex sets, or more generally, nonconvex functions. Therefore, when one wishes to use importance sampling to estimate the price of an option with nonconvex payoff (e.g., an outperformance option), exercise caution because selecting an alternative sampling density from the exponential family may lead to erroneous estimates. For a general approach to building efficient importance sampling schemes, see [9, 10].

Chapter 8

Stochastic Calculus

Stochastic calculus is an essential mathematical tool in modern continuous time finance [1, 7, 16, 19, 23]. For example, stochastic integrals naturally arise in the analysis of self-financing portfolios and Itô formula connects option prices with solutions to partial differential equations. Recent years have also seen the development of many financial models where the underlying stock prices are described by diffusion processes or solutions to stochastic differential equations. They are generalizations of the classical Black–Scholes model, which uses geometric Brownian motions to model stock prices.

Stochastic calculus differs fundamentally from classical calculus. While classical calculus often involves integrals against functions that are nice and smooth, stochastic calculus has to deal with integrals with respect to random processes whose sample paths are rugged and nowhere differentiable. As a consequence, Itô formula, instead of the fundamental theorem of classical calculus, becomes the corner stone of stochastic calculus.

This chapter offers a quick introduction to stochastic calculus, including stochastic integrals, Itô formula, and stochastic differential equations. Since a mathematically rigorous treatment is beyond the scope of this book, we aim for an informal understanding of stochastic calculus. This means that, for example, some technical conditions such as measurability and integrability will be omitted in the statement of theorems (they are satisfied in nearly all practical applications), the modes of convergence will not be clearly specified, solutions to stochastic differential equations will not be distinguished as in the strong or weak sense, and so on. For the more mathematically inclined reader, rigorous treatment of stochastic calculus can be found in many advanced graduate level textbooks such as [18, 25, 26, 27].

8.1   Stochastic Integrals

Stochastic integrals are integrals with respect to random processes. The most basic stochastic integral is an integral against a standard Brownian motion that takes the form

$$\int_0^T X_t \, dW_t, \tag{8.1}$$

where W is a standard Brownian motion and the integrand X is a general stochastic process. There are mild conditions one needs to impose on the integrand X so as to make the integral meaningful; see Remark 8.1. Throughout the rest of the book, these conditions are implicitly assumed.

Analogous to a classical integral, a stochastic integral is defined as the limit of certain approximating sums. More precisely, the stochastic integral (8.1) is defined as

$$\int_0^T X_t \, dW_t = \lim_{n \to \infty} \sum_{i=0}^{n-1} X_{t_i} (W_{t_{i+1}} - W_{t_i}), \tag{8.2}$$

where $0 = t_0 < t_1 < \cdots < t_n = T$ is a partition of the interval $[0, T]$. It should be mentioned that we always assume

$$\max_{0 \le i \le n-1} (t_{i+1} - t_i) \to 0$$

as n tends to infinity for any partition. The choice of partition does not affect the limit in (8.2). However, it is often convenient to let $t_i = iT/n$ for each i to ease computation.

We list below some of the fundamental properties of stochastic integrals. While linearity is obvious from the definition, the martingale property and Itô isometry are less so; see Remark 8.2 for a formal argument.

1. **Linearity:** Given any $a, b \in \mathbb{R}$,

$$\int_0^T (a X_t + b Y_t) \, dW_t = a \int_0^T X_t \, dW_t + b \int_0^T Y_t \, dW_t.$$

2. **Martingale property:**

$$E \int_0^T X_t \, dW_t = 0.$$

3. **Itô isometry:**

$$E\left(\int_0^T X_t\, dW_t\right)^2 = \int_0^T E[X_t^2]\, dt.$$

Another useful property is that when the integrand is deterministic, the stochastic integral is normally distributed. More precisely, we have the following lemma.

Lemma 8.1. *Assume that $f(t)$ is a deterministic function. Then the stochastic integral*

$$\int_0^T f(t)\, dW_t$$

is normally distributed with mean 0 and variance $\int_0^T f^2(t)\, dt$.

PROOF: By definition, this stochastic integral equals the limit of the approximating sum

$$S_n = \sum_{i=1}^{n-1} f(t_i)[W_{t_{i+1}} - W_{t_i}]$$

as n tends to infinity. It is easy to see that S_n is normally distributed with mean 0 and variance

$$\operatorname{Var}[S_n] = \sum_{i=1}^{n-1} f^2(t_i)[t_{i+1} - t_i].$$

Clearly,

$$\operatorname{Var}[S_n] \to \int_0^T f^2(t)\, dt$$

as n tends to infinity. The lemma follows readily. ∎

Example 8.1. The purpose of this example is to illustrate the difference between stochastic integral and classical integral by explicitly computing

$$\int_0^T 2W_t\, dW_t.$$

Note that this integral is an exception rather than the norm as stochastic integrals can rarely be explicitly evaluated.

SOLUTION: Given any n, consider the partition $0 = t_0 < t_1 < \cdots < t_n = T$ with $t_i = iT/n$. For each $i = 0, 1, \ldots, n-1$, let

$$Z_{i+1} = \frac{W_{t_{i+1}} - W_{t_i}}{\sqrt{t_{i+1} - t_i}} = \frac{W_{t_{i+1}} - W_{t_i}}{\sqrt{T/n}}.$$

Then $Z_1, \ldots, Z_n$ are iid standard normal random variables. It follows from definition (8.2) and the strong law of large numbers that

$$
\begin{aligned}
&\int_0^T 2W_t \, dW_t \\
&= \lim_n \sum_{i=0}^{n-1} 2W_{t_i}(W_{t_{i+1}} - W_{t_i}) \\
&= \lim_n \sum_{i=0}^{n-1} \left[(W_{t_{i+1}} + W_{t_i}) - (W_{t_{i+1}} - W_{t_i})\right](W_{t_{i+1}} - W_{t_i}) \\
&= W_T^2 - \lim_n \sum_{i=0}^{n-1} (W_{t_{i+1}} - W_{t_i})^2 \\
&= W_T^2 - \lim_n \frac{T}{n} \sum_{i=1}^{n} Z_i^2 \\
&= W_T^2 - T.
\end{aligned}
$$

Note that if we were to calculate this stochastic integral using classical calculus, we would arrive at the *wrong* conclusion that

$$\int_0^T 2W_t \, dW_t = \int_0^T d(W_t^2) = W_T^2.$$

The difference is due to the nonzero quadratic variation of Brownian motion, that is,

$$\lim_{n \to \infty} \sum_{i=0}^{n-1} (W_{t_{i+1}} - W_{t_i})^2 = T \neq 0. \tag{8.3}$$

In a classical integral, the corresponding quadratic variation term is zero. ∎

Remark 8.1. In order to ensure the stochastic integral is well defined, some conditions must be imposed on the integrand X. For one, X has to be *adapted* or *nonanticipating*. This condition basically says that the value of X_t only depends on the information up until time t. In practice, if X_t represents certain decisions an investor makes at time t, the adaptedness of X

means that the investor cannot look into the future and can only make the decision based upon the historical information available to him. Another condition that is often imposed on X requires the expected value

$$E \int_0^T X_t^2 \, dt$$

to be finite. These two conditions hold automatically in nearly all applications.

Remark 8.2. The martingale property and Itô isometry can be easily understood if one observes that X_{t_i} is independent of $\Delta W_j = W_{t_{j+1}} - W_{t_j}$ for any $j \geq i$. Indeed, by the adaptedness of X, X_{t_i} only depends on the historical information up until time t_i. On the other hand, the Brownian motion increment ΔW_j is independent of the historical information up to time t_j. The independence between X_{t_i} and ΔW_j follows since $t_j \geq t_i$. Consequently,

$$E \int_0^T X_t \, dW_t = \lim_n \sum_{i=0}^{n-1} E[X_{t_i} \Delta W_i] = \lim_n \sum_{i=0}^{n-1} E[X_{t_i}] \cdot E[\Delta W_i] = 0,$$

which is the martingale property. By the same token, for every i and $j > i$,

$$E[X_{t_i} \Delta W_i]^2 = E[X_{t_i}^2] \cdot E[\Delta W_i]^2 = E[X_{t_i}^2](t_{i+1} - t_i),$$

$$E[X_{t_i} \Delta W_i \cdot X_{t_j} \Delta W_j] = E[X_{t_i} X_{t_j} \Delta W_i] \cdot E[\Delta W_j] = 0.$$

Therefore,

$$
\begin{aligned}
E \left(\int_0^T X_t \, dW_t \right)^2 &= \lim_n E \left[\sum_{i=0}^{n-1} X_{t_i} \Delta W_i \right]^2 \\
&= \lim_n \sum_{i=0}^{n-1} E[X_{t_i}^2](t_{i+1} - t_i) \\
&= \int_0^T E[X_t^2] \, dt,
\end{aligned}
$$

which is exactly the Itô isometry.

Remark 8.3. The stochastic integral defined by (8.2) is said to be the *Itô integral*. The characteristic of Itô integral is that in the approximating sum the integrand X is always evaluated at the left endpoint of each of the subintervals $[t_i, t_{i+1}]$. If we replace it by the right endpoint $X_{t_{i+1}}$ or the average $(X_{t_i} + X_{t_{i+1}})/2$, the limit in (8.2) will be different and the resulting integral is said to be the *backward Itô integral* or the *Stratonovich integral*, respectively. This is very different from the classical Riemann integral, where the integrand can be evaluated at any point of the subintervals without affecting the limit. The reason for this difference is again that the Brownian motion has nonzero quadratic variation (8.3). In financial engineering, stochastic integrals are predominantly Itô integrals.

8.2   Itô Formula

The most significant result in stochastic calculus is without any doubt the celebrated *Itô formula*. Analogous to the chain rule in classical calculus, it characterizes the dynamics of functions of continuous time stochastic processes such as Brownian motions and diffusions. Itô formula plays a very important role in financial engineering. For instance, since the value of a financial derivative is in general a smooth function of the underlying asset price, Itô formula is applicable and eventually connects the value of the derivative with the solution to a partial differential equation. We will briefly touch upon this topic at the end of this chapter.

8.2.1   The Basic Itô Formula

The basic Itô formula states that a smooth function of a standard Brownian motion is the summation of a stochastic integral and an ordinary integral. It will be subsumed by the more general Itô formulas that will appear later. However, it is probably the simplest setting to illustrate the key difference between stochastic calculus and classical calculus.

Theorem 8.2. *Suppose that* $f : \mathbb{R} \to \mathbb{R}$ *is a twice continuously differentiable function and* W *is a standard Brownian motion. Then*

$$f(W_T) = f(W_0) + \int_0^T f'(W_t) \, dW_t + \frac{1}{2} \int_0^T f''(W_t) \, dt.$$

More general versions of Itô formula usually require less stringent regularity conditions on f and can accommodate a wider class of process models other than the standard Brownian motion. However, the characteristics remain the same. That is, compared with the classical chain rule, there should be an extra term analogous to

$$\frac{1}{2} \int_0^T f''(W_t)\, dt.$$

The intrinsic reason for the presence of this term is that the quadratic variation of a Brownian motion sample path is T over any time interval $[0, T]$; see equations (8.3) and (8.4).

"PROOF" OF THEOREM 8.2. In order to understand the idea behind Itô formula, we should present a formal and sketchy argument. For clarity, we will further assume that f'' is bounded by some constant M. This assumption is nonessential and can be removed by a more detailed analysis. Let $t_i = iT/n, i = 0, 1, \ldots, n,$ be a partition of the time interval $[0, T]$. We should eventually send n to infinity. Write

$$f(W_T) = f(W_0) + \sum_{i=0}^{n-1} \left[f(W_{t_{i+1}}) - f(W_{t_i}) \right].$$

Denote $\Delta W_i = W_{t_{i+1}} - W_{t_i}$ and $\Delta t_i = t_{i+1} - t_i$. It follows from the Taylor expansion that

$$f(W_{t_{i+1}}) - f(W_{t_i}) = f'(W_{t_i})\Delta W_i + \frac{1}{2}f''(\xi_i)(\Delta W_i)^2,$$

where ξ_i is a random variable taking values between W_{t_i} and $W_{t_{i+1}}$. By the definition (8.2) of stochastic integral,

$$\sum_{i=0}^{n-1} f'(W_{t_i})\Delta W_i \rightarrow \int_0^T f'(W_t)\, dW_t.$$

Furthermore, thanks to equation (8.3) and the continuity of f'' and W,

$$\sum_{i=0}^{n-1} |f''(\xi_i) - f''(W_{t_i})|(\Delta W_i)^2 \leq \max_i |f''(\xi_i) - f''(W_{t_i})| \sum_{i=0}^{n-1} (\Delta W_i)^2 \rightarrow 0.$$

Therefore, it suffices to show

$$\sum_{i=0}^{n-1} f''(W_{t_i})(\Delta W_i)^2 \rightarrow \int_0^T f''(W_t)\, dt. \tag{8.4}$$

The left-hand-side equals

$$\sum_{i=0}^{n-1} f''(W_{t_i})\Delta t_i + \sum_{i=0}^{n-1} f''(W_{t_i})[(\Delta W_i)^2 - \Delta t_i] = (\mathrm{I}) + (\mathrm{II}).$$

It is trivial that (I) converges to the right-hand-side of (8.4). Therefore, we only need to show (II) $\to$ 0. Note that W_{t_i} and ΔW_j are independent if $j \geq i$, therefore

$$
\begin{aligned}
E(\mathrm{II}) &= \sum_{i=0}^{n-1} E[f''(W_{t_i})] \cdot E[(\Delta W_i)^2 - \Delta t_i] = 0, \\
\mathrm{Var}(\mathrm{II}) &= \sum_{i=0}^{n-1} E\left\{ f''(W_{t_i})[(\Delta W_i)^2 - \Delta t_i] \right\}^2 \\
&\leq M^2 \cdot \sum_{i=0}^{n-1} E[(\Delta W_i)^2 - \Delta t_i]^2 \\
&= M^2 \cdot \sum_{i=0}^{n-1} (\Delta t_i)^2 E(Z_i^2 - 1)^2 \\
&= M^2 \cdot E(Z^2 - 1)^2 \cdot \frac{T^2}{n} \to 0,
\end{aligned}
$$

where Z and Z_i's are iid standard normal random variables. It follows that (II) converges to zero. We complete the argument. ∎

Remark 8.4. Loosely speaking, since $(\Delta W_i)^2 = Z_i^2 \Delta t_i$ for some standard normal random variable Z_i, ΔW_i is of order $\sqrt{\Delta t_i}$. This leads to the non-vanishing limit (8.4), which is precisely the reason for the extra term in Itô formula. It also explains why we have applied the Taylor expansion up until order 2, that is, the higher order terms in the Taylor expansion are of orders higher than Δt_i.

8.2.2   The Differential Notation

It is notationally cumbersome to express Itô formula in its integral form. In literature, one often adopts the so-called *differential notation*. More precisely, the differential notation for a process X that satisfies

$$
X_t = X_0 + \int_0^t Y_s \, ds + \int_0^t Z_s \, dW_s
$$

is simply

$$
dX_t = Y_t \, dt + Z_t \, dW_t. \tag{8.5}
$$

Even though the differential notation has been used extensively, one should keep in mind that it is only a notational convenience. The true meaning always lies with the integral form.

In stochastic calculus, it is also convenient to define the *product* of two differentials in such a way that

$$(dt)^2 = 0, \quad dt\, dW_t = dW_t dt = 0, \quad (dW_t)^2 = dt$$

for any standard Brownian motion W. Not surprisingly, $(dW_t)^2 = dt$ reflects the fact that the quadratic variation of a standard Brownian motion sample path is T over any time interval $[0, T]$. From these definitions one can easily calculate the product of general differentials. For example, if

$$
\begin{aligned}
dX_t &= Y_t\, dt + Z_t\, dW_t \\
dM_t &= R_t\, dt + Q_t\, dW_t,
\end{aligned}
$$

then

$$
\begin{aligned}
dX_t dM_t &= Y_t R_t (dt)^2 + (Y_t Q_t + Z_t R_t)(dt dW_t) + Z_t Q_t (dW_t)^2 \\
&= Z_t Q_t\, dt.
\end{aligned}
$$

When W and B are two independent standard Brownian motions, the product of their differentials is defined to be

$$dW_t dB_t = 0.$$

See also Exercise 8.10.

With this notation, Theorem 8.2 can be written in a much more succinct fashion

$$df(W_t) = f'(W_t)\, dW_t + \frac{1}{2} f''(W_t)\, (dW_t)^2,$$

which is very easy to memorize because of its resemblance to the Taylor expansion.

8.2.3 General Itô Formulas and Product Rule

We state a general Itô formula below. The proof is beyond the scope of this book and thus omitted. However, the main idea is exactly the same as in Theorem 8.2. Throughout the section, all relevant processes are assumed to be of the form (8.5).

Theorem 8.3. *Suppose that $f(t, x) : [0, \infty) \times \mathbb{R}^d \to \mathbb{R}$ is continuously differentiable with respect to t and twice continuously differentiable with respect to x.*

Then for a d-dimensional process $X_t = (X_t^{(1)}, \ldots, X_t^{(d)})$,

$$df(t, X_t) = \frac{\partial f}{\partial t}(t, X_t)\, dt + \sum_{i=1}^{d} \frac{\partial f}{\partial x_i}(t, X_t)\, dX_t^{(i)}$$

$$+ \frac{1}{2} \sum_{i=1}^{d} \sum_{j=1}^{d} \frac{\partial^2 f}{\partial x_i \partial x_j}(t, X_t)\, dX_t^{(i)}\, dX_t^{(j)}.$$

We also want to mention two very useful corollaries. Letting $d = 1$ and f be a function of x alone, we arrive at Corollary 8.4, of which the basic Itô formula in Theorem 8.2 is a special case with $X = W$. On the other hand, if we let $d = 2$ and $f(t, x, y) = xy$, then the product rule follows.

Corollary 8.4. *Suppose that* $f : \mathbb{R} \to \mathbb{R}$ *is twice continuously differentiable. Then*

$$df(X_t) = f'(X_t)\, dX_t + \frac{1}{2} f''(X_t)(dX_t)^2.$$

Corollary 8.5. Product Rule. $d(X_t Y_t) = X_t\, dY_t + Y_t\, dX_t + dX_t\, dY_t.$

Example 8.2. Let W be a standard Brownian motion. Consider a geometric Brownian motion

$$S_t = S_0 \exp\left\{ \left(r - \frac{1}{2}\sigma^2 \right) t + \sigma W_t \right\}.$$

Compute the differential dS_t.

SOLUTION: This is a straightforward application of Itô formula. Observe that $S_t = f(t, W_t)$, where

$$f(t, x) = S_0 \exp\left\{ \left(r - \frac{1}{2}\sigma^2 \right) t + \sigma x \right\}.$$

It follows from Theorem 8.3 and direct calculation (we omit the details) that

$$\begin{aligned} dS_t &= \frac{\partial f}{\partial t}(t, W_t)dt + \frac{\partial f}{\partial x}(t, W_t)dW_t + \frac{1}{2}\frac{\partial^2 f}{\partial x^2}(t, W_t)(dW_t)^2 \\ &= rS_t\, dt + \sigma S_t\, dW_t. \end{aligned}$$

This is indeed the formula often seen in the mathematical finance literature for asset prices that are geometric Brownian motions. ∎

Example 8.3. Suppose that $dS_t = Y_t\,dt + Z_t\,dW_t$ and f is a twice continuously differentiable function. Define

$$X_t = e^{-rt}f(S_t).$$

Compute the differential dX_t.

SOLUTION: Let $R_t = e^{-rt}$ and $H_t = f(S_t)$. Then it follows from Corollary 8.4 that

$$
\begin{aligned}
dH_t &= f'(S_t)\,dS_t + \frac{1}{2}f''(S_t)\,(dS_t)^2 \\
&= f'(S_t)(Y_t\,dt + Z_t\,dW_t) + \frac{1}{2}f''(S_t)Z_t^2\,dt.
\end{aligned}
$$

It is trivial that $dR_t = -rR_t\,dt$. By the product rule (note that $dR_t\,dH_t = 0$),

$$
\begin{aligned}
dX_t &= R_t\,dH_t + H_t\,dR_t + dR_t\,dH_t \\
&= e^{-rt}\left[-rf(S_t) + f'(S_t)Y_t + \frac{1}{2}f''(S_t)Z_t^2\right]dt + e^{-rt}f'(S_t)Z_t\,dW_t.
\end{aligned}
$$

Another way to calculate dX_t is to write $X_t = g(t, S_t)$ where $g(t, x) = e^{-rt}f(x)$ and then apply the general Itô formula in Theorem 8.3. It leads to, of course, exactly the same result. ∎

Example 8.4. Let W be a standard Brownian motion and Y an arbitrary process. Define

$$
\begin{aligned}
M_t &= \exp\left\{\int_0^t Y_s\,dW_s - \frac{1}{2}\int_0^t Y_s^2\,ds\right\}, \\
X_t &= W_t - \int_0^t Y_s\,ds.
\end{aligned}
$$

Compute the differentials dM_t and $d(X_tM_t)$.

SOLUTION: It is often helpful to introduce some intermediate processes to facilitate the calculation of differentials. Let

$$Q_t = \int_0^t Y_s\,dW_s - \frac{1}{2}\int_0^t Y_s^2\,ds.$$

Then $M_t = e^{Q_t}$ and

$$dQ_t = Y_t\,dW_t - \frac{1}{2}Y_t^2\,dt, \quad (dQ_t)^2 = Y_t^2\,dt.$$

It follows from Corollary 8.4 that

$$dM_t = e^{Q_t} dQ_t + \frac{1}{2} e^{Q_t} (dQ_t)^2 = M_t \left[dQ_t + \frac{1}{2} (dQ_t)^2 \right] = M_t Y_t \, dW_t.$$

To calculate $d(X_t M_t)$, one can just use the product rule:

$$
\begin{aligned}
d(X_t M_t) &= X_t \, dM_t + M_t \, dX_t + dX_t \, dM_t \\
&= X_t M_t Y_t \, dW_t + M_t (dW_t - Y_t \, dt) + M_t Y_t dt \\
&= (X_t M_t Y_t + M_t) \, dW_t.
\end{aligned}
$$

The explicit calculation of the differentials has some interesting implications. For example, the form of dM_t implies that

$$M_t = M_0 + \int_0^t M_s Y_s \, dW_s = 1 + \int_0^t M_s Y_s \, dW_s.$$

Thanks to the martingale property of stochastic integrals, it follows that $E[M_t] = 1$ (this is not entirely accurate because some conditions on Y are needed to ensure the martingale property). Further analysis will eventually lead to the famous Girsanov's Theorem [18]. ∎

8.3   Stochastic Differential Equations

In many financial models, the price of the underlying asset is assumed to be a *diffusion process*, that is, the solution to a *stochastic differential equation* of the following form:

$$dX_t = b(t, X_t) \, dt + \sigma(t, X_t) \, dW_t, \tag{8.6}$$

where b (drift) and σ (volatility) are some given function. We say X is a solution to the stochastic differential equation (8.6) if it satisfies

$$X_T = X_0 + \int_0^T b(t, X_t) \, dt + \int_0^T \sigma(t, X_t) \, dW_t$$

for every T and some standard Brownian motion W; see Remark 8.5. Very few stochastic differential equations admit explicit solutions. However, they are not difficult to approximate by Monte Carlo simulation.

Example 8.5. Let r be a constant and $\theta(t)$ an arbitrarily given deterministic function. Solve the stochastic differential equation

$$dX_t = rX_t\, dt + \theta(t)\, dW_t, \quad X_0 = x.$$

SOLUTION: Consider the process $Y_t = e^{-rt}X_t$. Then $Y_0 = X_0 = x$. It follows from the product rule that

$$dY_t = -re^{-rt}X_t\, dt + e^{-rt}\, dX_t = e^{-rt}\theta(t)\, dW_t.$$

Therefore,

$$X_T = e^{rT}Y_T = e^{rT}\left(x + \int_0^T e^{-rt}\theta(t)\, dW_t \right).$$

Even though this is not exactly an explicit formula because of the stochastic integral involved, the distribution of X_T is

$$N\left(e^{rT}x,\ e^{2rT}\int_0^T e^{-2rt}\theta^2(t)\, dt \right)$$

from Lemma 8.1. This is all one needs to simulate samples of X_T. ∎

Example 8.6. Let a, b, and σ be arbitrary constants. Solve the stochastic differential equation

$$dX_t = a(b - X_t)\, dt + \sigma\, dW_t, \quad X_0 = x.$$

SOLUTION: Analogous to the previous example, we define $Y_t = e^{at}X_t$ and use the product rule to obtain

$$dY_t = ae^{at}X_t\, dt + e^{at}\, dX_t = abe^{at}\, dt + \sigma e^{at}\, dW_t,$$

which implies that

$$\begin{aligned}
X_T &= e^{-aT}Y_T \\
&= e^{-aT}\left(x + \int_0^T abe^{at}\, dt + \int_0^T \sigma e^{at}\, dW_t \right) \\
&= e^{-aT}x + b(1 - e^{-aT}) + e^{-aT}\int_0^T \sigma e^{at}\, dW_t.
\end{aligned}$$

Thanks to Lemma 8.1, X_T is normally distributed as

$$N\left(e^{-aT}x + b(1 - e^{-aT}),\ \frac{\sigma^2}{2a}(1 - e^{-2aT}) \right).$$

It is particularly interesting when a is positive. In this case, X is said to be a *mean-reverting* Ornstein–Uhlenbeck process. It is mean-reverting because if $X_t > b$, then the drift $a(b - X_t)$ is negative and pushes X_t down; on the other hand, if $X_t < b$, then the drift is positive and drives the process up. Therefore, X oscillates around b. Coincidentally, the limit distribution of X_T as $T \to \infty$ is

$$N\left(b, \frac{\sigma^2}{2a}\right),$$

whose mean is exactly b.   ■

Example 8.7. Let r be a constant and $\theta(t)$ an arbitrarily given deterministic function. Solve the stochastic differential equation

$$dX_t = rX_t\,dt + \theta(t)X_t\,dW_t, \quad X_0 = x > 0.$$

SOLUTION: Define the process $Y_t = \log X_t$. Then $Y_0 = \log X_0 = \log x$. It follows from Corollary 8.4 that

$$dY_t = \frac{1}{X_t}\,dX_t - \frac{1}{2}\frac{1}{X_t^2}(dX_t)^2 = r\,dt + \theta(t)\,dW_t - \frac{1}{2}\theta^2(t)\,dt.$$

Therefore,

$$X_T = e^{Y_T} = x \cdot \exp\left\{\int_0^T \left(r - \frac{1}{2}\theta^2(t)\right)dt + \int_0^T \theta(t)\,dW_t\right\}.$$

Letting $\theta(t)$ be a constant, say σ, we recover the classical geometric Brownian motion.   ■

Continuous Time Financial Models: For the purpose of demonstration, we list a small collection of financial models that involve diffusion processes. Most of these models do not admit explicit solutions.

a. **Black–Scholes model:** The risk-free interest rate is assumed to be a constant r. The stock price is a geometric Brownian motion with drift μ and volatility σ:

$$\frac{dS_t}{S_t} = \mu\,dt + \sigma\,dW_t.$$

b. **CEV model:** The stock price under the *constant elasticity of variance* model is assumed to satisfy

$$dS_t = \mu S_t\,dt + \sigma S_t^\gamma\,dW_t,$$

where $\gamma > 0$ is a constant.

c. **Stochastic volatility models:** The volatility of the underlying asset is a diffusion process itself. One of the most widely used stochastic volatility models is the Heston model [15]:

$$\frac{dS_t}{S_t} = \mu \, dt + \sqrt{\theta_t} \, dW_t,$$
$$d\theta_t = a(b - \theta_t) \, dt + \sigma \sqrt{\theta_t} \, dB_t,$$

where $a, b > 0$ and (W, B) is a two-dimensional Brownian motion with covariance matrix

$$\begin{bmatrix} 1 & \rho \\ \rho & 1 \end{bmatrix}.$$

d. **Interest rate models:** In the context of pricing interest rate derivatives, modeling the *instantaneous interest rate* or *short rate* is an important issue. Below are a couple of such models. The short rate at time t is denoted by r_t.

$$\text{(Vasicek)} \qquad dr_t = a(b - r_t) \, dt + \sigma \, dW_t.$$
$$\text{(Cox–Ingersoll–Ross)} \qquad dr_t = a(b - r_t) \, dt + \sigma \sqrt{r_t} \, dW_t.$$

Remark 8.5. Rigorously speaking, there should be two types of solutions to a stochastic differential equation: *strong solutions* and *weak solutions*. The key difference between these two notions is whether the driving Brownian motion W is designated as a given input (strong solution) or viewed as part of the solution itself (weak solution). The concept of weak solutions suffices in nearly all financial applications for the following reasons. (i) In financial modeling, what really matters is the distributional properties of the relevant processes. To identify the driving Brownian motion W is both difficult and unnecessary. (ii) The condition for the existence of a strong solution is much more stringent than that of a weak solution.

8.4 Risk-Neutral Pricing

The option pricing formula for the classical Black–Scholes model can be generalized to models with time varying (deterministic or stochastic) interest rate and complex volatility structure. To be more concrete, denote by r_t the instantaneous interest rate or short rate at time t. Then \$1 at time 0 is worth

$$\exp\left\{ \int_0^T r_t \, dt \right\}$$

at time T. It is not surprising that the price of an option with maturity T and payoff X should be

$$v = E \left[\exp \left\{ - \int_0^T r_t \, dt \right\} X \right], \tag{8.7}$$

where the expected value is taken with respect to the risk-neutral probability measure. The underlying asset price can be diffusion processes other than the geometric Brownian motion. However, the drift must equal the short rate r_t under the risk-neutral probability measure. An example of such a price process is

$$\frac{dS_t}{S_t} = r_t \, dt + \theta(t, S_t) \, dW_t,$$

where θ is some given function and W is a standard Brownian motion. In the special case where $r_t \equiv r$ and $\theta(t, x) \equiv \sigma$, we recover the classical pricing formula for the Black–Scholes model.

Example 8.8. Assume that the risk-free interest rate r is a constant and the price of the underlying stock satisfies the stochastic differential equation

$$\frac{dS_t}{S_t} = r \, dt + \theta(t) \, dW_t$$

for some deterministic function $\theta(t)$ under the risk-neutral probability measure. Find the price of a call option with maturity T and strike price K.

SOLUTION: The price of the option is $v = E[e^{-rT}(S_T - K)^+]$. Thanks to Example 8.7 and Lemma 8.1,

$$\begin{aligned} S_T &= S_0 \exp \left\{ \int_0^T \left(r - \frac{1}{2}\theta^2(t) \right) dt + \int_0^T \theta(t) \, dW_t \right\} \\ &= S_0 \exp \left\{ \int_0^T \left(r - \frac{1}{2}\sigma^2 \right) dt + \sigma\sqrt{T}Z \right\}, \end{aligned}$$

where Z is a standard normal random variable and

$$\sigma = \sqrt{\frac{1}{T} \int_0^T \theta^2(t) \, dt}.$$

In other words, S_T has the same distribution as the terminal stock price in the classical Black–Scholes model with drift r and volatility σ. It follows that $v = \text{BLS\_Call}(S_0, K, T, r, \sigma)$.   ■

Example 8.9. Consider the Ho–Lee model where the short rate r_t satisfies the stochastic differential equation

$$dr_t = \theta(t)\, dt + \sigma\, dW_t$$

under the risk-neutral probability measure. Assuming that $\theta(t)$ is a deterministic function, compute the price of a zero-coupon bond with payoff \$1 and maturity T.

SOLUTION: The price of this zero-coupon bond is given by the pricing formula (8.7) with $X = 1$, that is,

$$v = E\left[\exp\left\{-\int_0^T r_t\, dt\right\}\right].$$

Observe that

$$\int_0^T r_t\, dt = \int_0^T \left[r_0 + \int_0^t \theta(s)\, ds + \sigma W_t\right] dt = \mu + \sigma Y,$$

where

$$\mu = r_0 T + \int_0^T\!\!\int_0^t \theta(s)\, ds\, dt, \quad Y = \int_0^T W_t\, dt.$$

In order to determine the distribution of Y, consider an arbitrary partition $0 = t_0 < t_1 < \cdots < t_n = T$. Letting $\Delta t_i = t_{i+1} - t_i$, Y is the limit of

$$S_n = \sum_{i=0}^{n-1} W_{t_i} \Delta t_i.$$

For each n, S_n is normally distributed with mean 0. It follows from Lemma 1.7 and Exercise 2.6 that

$$
\begin{aligned}
\mathrm{Var}[S_n] &= \sum_{i=0}^{n-1} \mathrm{Var}[W_{t_i}](\Delta t_i)^2 + 2\sum_{i<j} \mathrm{Cov}[W_{t_i}, W_{t_j}]\Delta t_i \Delta t_j \\
&= \sum_{i=0}^{n-1} t_i (\Delta t_i)^2 + 2\sum_{i<j} t_i \Delta t_i \Delta t_j \\
&= \sum_{i=0}^{n-1} t_i (\Delta t_i)^2 + 2\sum_{i=0}^{n-1} (T - t_{i+1}) t_i \Delta t_i \\
&= -\sum_{i=0}^{n-1} t_i (\Delta t_i)^2 + 2\sum_{i=0}^{n-1} (T - t_i) t_i \Delta t_i.
\end{aligned}
$$

Therefore, Y is normally distributed itself with mean 0 and variance

$$\beta^2 = \text{Var}[Y] = \lim_{n \to \infty} \text{Var}[S_n] = 0 + 2 \int_0^T (T - t)t \, dt = \frac{1}{3} T^3,$$

and the bond price is

$$v = E\left[e^{-\mu - \sigma Y}\right] = e^{-\mu + \frac{1}{2}\sigma^2 \beta^2}.$$

See Exercise 8.8 for a different approach to calculate $\text{Var}[Y]$.   ■

8.5   Black–Scholes Equation

The most famous application of Itô formula to the option pricing theory is probably about the connection between option prices and a family of second order partial differential equations. This leads to an entirely different approach to option pricing by means of partial differential equations. Even though the focus of the book is about Monte Carlo simulation, it is worthwhile to briefly discuss this connection.

Consider a call option with strike price K and maturity T in the classical Black–Scholes model, where the stock price is a geometric Brownian motion with drift r and volatility σ under the risk-neutral probability measure:

$$dS_t = rS_t \, dt + \sigma S_t \, dW_t.$$

Define for every $t \in [0, T]$ and $x > 0$

$$v(t, x) = E[e^{-r(T-t)} (S_T - K)^+ | S_t = x].$$

That is, $v(t, x)$ is the price of the call option evaluated at time t if the stock price at time t is x. By definition, the call option price at time 0 is $v(0, S_0)$. Moreover, for any $\theta \in [0, T]$ it follows from the tower property that

$$v(0, S_0) = E[e^{-rT} (S_T - K)^+] = E[e^{-r\theta} v(\theta, S_\theta)].$$

Assuming that v is nice and smooth, Itô formula implies that

$$e^{-r\theta} v(\theta, S_\theta) = v(0, S_0) + \int_0^\theta e^{-rt} \mathbb{L}v(t, S_t) \, dt + \int_0^\theta e^{-rt} \frac{\partial v}{\partial x}(t, S_t) \sigma S_t \, dW_t,$$

where

$$\mathbb{L}v = -rv + \frac{\partial v}{\partial t} + \frac{\partial v}{\partial x} rx + \frac{1}{2}\sigma^2 x^2 \frac{\partial^2 v}{\partial x^2}.$$

Taking expected value on both sides and observing that by the martingale property the expected value of the stochastic integral is zero, we have

$$E[e^{-r\theta}v(\theta, S_\theta)] = v(0, S_0) + \int_0^\theta e^{-rt} E[\mathbb{L}v(t, S_t)]\, dt.$$

Therefore,

$$\int_0^\theta e^{-rt} E[\mathbb{L}v(t, S_t)]\, dt = 0$$

for every θ, which in turn implies that $E[\mathbb{L}v(t, S_t)] = 0$ for every t. Now letting $t \to 0$, we arrive at $\mathbb{L}v(0, S_0) = 0$. However, in the above derivation, the initial time and initial stock price can be *arbitrary*, which leads to $\mathbb{L}v(t, x) = 0$ for every $t \in [0, T)$ and $x > 0$. In other words, v satisfies the partial differential equation

$$-rv + \frac{\partial v}{\partial t} + \frac{\partial v}{\partial x}rx + \frac{1}{2}\sigma^2 x^2 \frac{\partial^2 v}{\partial x^2} = 0, \quad 0 \le t < T,\, x > 0, \qquad (8.8)$$

with terminal condition (which is trivial from the definition of v)

$$v(T, x) = (x - K)^+, \quad x > 0. \qquad (8.9)$$

Equation (8.8) is called the *Black–Scholes equation*. Solving (8.8)–(8.9) yields the classical Black–Scholes formula of call option price.

More generally, if the option payoff is a function of the terminal stock price, say $h(S_T)$, then the option price satisfies the same Black–Scholes equation (8.8) but with a different terminal condition

$$v(T, x) = h(x).$$

This can be easily verified by repeating the preceding derivation. The situation is also very similar if the stock price is other than a geometric Brownian motion; see Exercise 8.16. More discussions on option pricing by means of partial differential equations can be found in the introductory textbook [30] or the more advanced [16, 19].

Exercises

8.1 Recover the result in Example 8.1 using the basic Itô formula with $f(x) = x^2$.

8.2 Use Itô formula to compute dX_t.

(a) $X_t = W_t^3$.

(b) $X_t = e^t W_t$.

(c) $X_t = W_t \int_0^t W_s \, ds$.

(d) $X_t = W_t \int_0^t s \, dW_s$.

8.3 Suppose that $dX_t = W_t \, dt + dW_t$. Let $Y_t = X_t^2$. Compute dY_t and $d(t^2 Y_t)$.

8.4 Define $X_t = W_t \exp\{-rt + \int_0^T Y_t \, dW_t\}$, where Y is a stochastic process and r is a constant. Compute dX_t.

8.5 Suppose that $W = (W^{(1)}, \ldots, W^{(d)})$ is a d-dimensional standard Brownian motion with $d \geq 2$. Let

$$R_t = \sqrt{(W_t^{(1)})^2 + \cdots + (W_t^{(d)})^2}.$$

Compute dR_t. R is said to be a *Bessel Process*.

8.6 Find the distribution of $\exp\{\int_0^T t \, dW_t\}$.

8.7 Find the distribution of X_T.

(a) $dX_t = dt + t \, dW_t$ and $X_0 = x$.

(b) $dX_t = -aX_t \, dt + dW_t$ and $X_0 = x$.

8.8 Here is another approach to determine the variance of Y in Example 8.9. Show that

$$\text{Var}[Y] = E[Y^2] = E\left[\int_0^T W_s \, ds \int_0^T W_t \, dt\right] = \int_0^T \int_0^T E[W_s W_t] \, ds \, dt.$$

Compute the double integral to obtain $\text{Var}[Y]$. Use this approach to find the distribution of $\int_0^T \theta(t) W_t \, dt$ for a given deterministic function $\theta(t)$.

8.9 Let $\theta(t)$ and $\sigma(t)$ be two deterministic functions. Show that the random vector

$$\left(\int_0^T \theta(t) \, dW_t, \int_0^T \sigma(t) \, dW_t\right)$$

is jointly normal with mean zero and covariance matrix $\Sigma = [\Sigma_{ij}]$ where

$$\Sigma_{11} = \int_0^T \theta^2(t) \, dt, \quad \Sigma_{12} = \Sigma_{21} = \int_0^T \theta(t)\sigma(t) \, dt, \quad \Sigma_{22} = \int_0^T \sigma^2(t) \, dt.$$

Hint: Given $0 = t_0 < t_1 < \cdots < t_m = T$, consider the approximating sums

$$X_n = \sum_{i=0}^{n-1} \theta(t_i)(W_{t_{i+1}} - W_{t_i}), \quad Y_n = \sum_{i=0}^{n-1} \sigma(t_i)(W_{t_{i+1}} - W_{t_i}).$$

Show that X_n and Y_n are jointly normal and compute their covariance.

8.10 Suppose that (W, B) is a two-dimensional Brownian motion with covariance matrix

$$\begin{bmatrix} 1 & \rho \\ \rho & 1 \end{bmatrix}$$

Show that $dW_t dB_t = \rho dt$. *Hint:* Cholesky factorization.

8.11 Let $\theta(t)$ be a deterministic function. Solve the stochastic differential equation

$$dX_t = \theta(t) dt + \sigma X_t dW_t, \quad X_0 = x.$$

Hint: Consider the process $X_t Y_t$ where $Y_t = \exp\left\{ \frac{1}{2}\sigma^2 t - \sigma W_t \right\}$.

8.12 Let a, b, and σ be constants. Solve the stochastic differential equation

$$dX_t = a(b - X_t) dt + \sigma X_t dW_t, \quad X_0 = x.$$

Hint: Let $Y_t = e^{at} X_t$. Derive the equation for Y.

8.13 Given two processes θ and σ, show that the solution to the stochastic differential equation $dX_t = \theta_t X_t dt + \sigma_t X_t dW_t$ is

$$X_T = X_0 \exp\left\{ \int_0^T \left(\theta_t - \frac{1}{2}\sigma_t^2 \right) dt + \int_0^T \sigma_t dW_t \right\}.$$

8.14 Assume that under the risk-neutral probability measure, the stock price satisfies the stochastic differential equation

$$dS_t = rS_t dt + \sigma(t, S_t) dW_t,$$

where r is the risk-free interest rate and $\sigma(t, x)$ is some function. Show that the *put-call parity* still holds. That is

$$v_c - v_p = S_0 - e^{-rT} K,$$

where v_c is the price of a call option with maturity T and strike price K, and v_p is that of a put option with the same maturity and strike price.

8.15 Suppose that $u(t, x) : [0, \infty) \times \mathbb{R} \to \mathbb{R}$ is a solution to the *heat equation*

$$\frac{\partial u}{\partial t} = \frac{1}{2} \frac{\partial^2 u}{\partial x^2}, \quad t > 0, \ x \in \mathbb{R}$$

with initial condition

$$u(0, x) = f(x), \quad x \in \mathbb{R}.$$

(a) Fix arbitrarily $T > 0$ and $x \in \mathbb{R}$. Define $X_t = x + W_t$ and $Y_t = u(T - t, X_t)$ for $t \in [0, T]$. Compute dY_t.

(b) Use the martingale property to argue that $E[Y_T] = E[Y_0] = u(T, x)$.

(c) Show that $E[Y_T] = E[f(X_T)] = E[f(x + W_T)]$. Therefore

$$u(T, x) = \int_{\mathbb{R}} f(x + y) \frac{1}{\sqrt{2\pi T}} e^{-y^2/(2T)} \, dy.$$

This type of argument is said to be a *verification argument*. The representation of the solution to a heat equation in terms of a Brownian motion is a special case of the so called *Feynman–Kac formula* [18].

8.16 Consider a call option with strike price K and maturity T. Assume that the risk-free interest rate r is a constant and the stock price S satisfies the stochastic differential equation

$$dS_t = rS_t \, dt + \theta(S_t) \, dW_t$$

under the risk-neutral probability measure. Let $v(t, x)$ be the value of the option at time t given that the stock price at time t is x. Argue that v satisfies the partial differential equation

$$-rv + \frac{\partial v}{\partial t} + rx \frac{\partial v}{\partial x} + \frac{1}{2} \theta^2(x) \frac{\partial^2 v}{\partial x^2} = 0, \quad 0 \le t < T, \ x > 0$$

with terminal condition

$$v(T, x) = (x - K)^+.$$

What should the partial differential equation be if the payoff of the option is $h(S_T)$ for some function h?

Chapter 9

Simulation of Diffusions

In general, diffusions or solutions to stochastic differential equations do not admit explicit formulas. To obtain quantitative information, one often resorts to Monte Carlo simulation. The idea is to discretize time and approximate the value of the diffusion at discrete time steps. Except for special cases, this operation will introduce *discretization error*, and thus the resulting estimates are usually biased. The variance reduction techniques we have discussed so far can also be brought into play. However, they will only reduce the variance of the estimate and will not affect the bias.

In this chapter, we state some of the most basic discretization schemes for stochastic differential equations and show how they can be combined with variance reduction techniques. Even though there are well-defined performance criteria to classify various discretization schemes, it is not our intention to do so here. The interested reader may want to consult the classical textbook [20] for further investigation.

9.1 Euler Scheme

Euler Scheme is the most intuitive and straightforward approximation to stochastic differential equations. To fix ideas, let $b(t, x)$ and $\sigma(t, x)$ be two continuous functions and consider the stochastic differential equation

$$dX_t = b(t, X_t)\, dt + \sigma(t, X_t)\, dW_t.$$

In order to simulate the sample paths of X on the time interval $[0, T]$, define a time discretization $0 = t_0 < t_1 < \cdots < t_m = T$. The approximating process $\hat{X}$ is defined recursively by

$$\hat{X}_0 = X_0, \tag{9.1}$$
$$\hat{X}_{t_{i+1}} = \hat{X}_{t_i} + b(t_i, \hat{X}_{t_i})(t_{i+1} - t_i) + \sigma(t_i, \hat{X}_{t_i})(W_{t_{i+1}} - W_{t_i})$$
$$= \hat{X}_{t_i} + b(t_i, \hat{X}_{t_i})(t_{i+1} - t_i) + \sigma(t_i, \hat{X}_{t_i})\sqrt{t_{i+1} - t_i}Z_{i+1}, \tag{9.2}$$

for $i = 0, \ldots, m - 1$, where $\{Z_1, \ldots, Z_m\}$ are iid standard normal random variables. In other words, in Euler scheme both the drift and the volatility are approximated by their values at the left-hand endpoint on each time interval $[t_i, t_{i+1})$.

Consider the problem of estimating $v = E[h(X_T)]$ for some function h. The Euler scheme will generate n iid sample paths and form the estimate

$$\hat{v} = \frac{1}{n} \sum_{i=1}^{n} h(\hat{X}_{T,i}),$$

where $\hat{X}_{T,1}, \ldots, \hat{X}_{T,n}$ denote the terminal values of these n sample paths. In general, since the distribution of $\hat{X}_T$ is different from that of X_T due to the time discretization,

$$E[\hat{v}] = E[h(\hat{X}_T)] \neq E[h(X_T)].$$

That is, the estimate $\hat{v}$ is *biased*. Moreover, this bias or discretization error cannot be eliminated by merely increasing the sample size n. Indeed, the strong law of large numbers implies that as n goes to infinity

$$\hat{v} - v \to E[h(\hat{X}_T)] - E[h(X_T)] \neq 0.$$

Note that variance reduction techniques will only accelerate the convergence of the estimate $\hat{v}$ to $E[h(\hat{X}_T)]$. They have no effect on the bias. On the other hand, it can be shown that under mild conditions the discretization error goes to zero as the time discretization becomes finer and finer, namely, as m tends to infinity. Given a computational budget, there is a trade off between increasing m for a smaller bias and increasing n for a smaller variance. Such analysis is often carried out under the principle of minimizing the mean square error of the estimate [8, 11].

Remark 9.1. Even though we have only described the Euler scheme for the one-dimensional diffusion processes, the extension to higher dimensions is trivial. Assume that X is a k-dimensional process and W is a d-dimensional Brownian motion with covariance matrix Σ. The Euler scheme (9.1)–(9.2) remains the same except that $\{Z_1, \ldots, Z_m\}$ become iid d-dimensional jointly normal random vectors with distribution $N(0, \Sigma)$.

9.2 Eliminating Discretization Error

The discretization error cannot be eliminated in general. However, for some special models it is possible. For example, consider a diffusion process X that satisfies the stochastic differential equation

$$dX_t = r\,dt + \theta(t)\,dW_t$$

for some constant r and deterministic function $\theta(t)$. Thanks to Lemma 8.1,

$$
\begin{aligned}
X_{t_{i+1}} - X_{t_i} &= r(t_{i+1} - t_i) + \int_{t_i}^{t_{i+1}} \theta(s)\,dW_s \\
&= r(t_{i+1} - t_i) + \sqrt{\int_{t_i}^{t_{i+1}} \theta^2(s)\,ds} \cdot Z_{i+1},
\end{aligned}
$$

where Z_{i+1} is a standard normal random variable. It leads to the following algorithm

$$\hat{X}_0 = X_0, \quad \hat{X}_{t_{i+1}} = \hat{X}_{t_i} + r(t_{i+1} - t_i) + \sqrt{\int_{t_i}^{t_{i+1}} \theta^2(s)\,ds} \cdot Z_{i+1},$$

where $\{Z_1, \ldots, Z_m\}$ are iid standard normal random variables. It is not difficult to see that $\hat{X}_{t_i}$ has the same distribution as X_{t_i} for every i, and thus there is no discretization error.

Actually, the time discretization is not really necessary if the quantity of interest only involves X_T since one can directly simulate X_T by

$$X_T = X_0 + rT + \int_0^T \theta(t)\,dW_t = X_0 + rT + \sqrt{\int_0^T \theta^2(t)\,dt} \cdot Z,$$

where Z is a standard normal random variable. However, time discretization is still needed in case one is interested in estimating an expected value of the more general form

$$E[h(X_0, X_{t_1}, \ldots, X_{t_{m-1}}, X_{t_m})],$$

or when X is brought into play as an auxiliary process (e.g., as a control variate) and the discretization is forced upon X.

It is worth noting that even if it is possible to simulate X_{t_i} exactly, discretization error may still exist if the quantity of interest can only be approximated. For instance, options that depend on the maximum or minimum of the underlying asset price such as lookback options, or the average of the underlying asset price such as Asian options, can only be approximated in general.

9.3   Refinements of Euler Scheme

There are various refinements of the plain Euler scheme. The goal of these refined schemes is to improve the approximation and reduce the discretization error. They are *not* variance reduction techniques. We will discuss one of such refinements under the assumption that X is a diffusion process of the form

$$dX_t = b(X_t)\, dt + \sigma(X_t)\, dW_t,$$

where b and σ are twice continuously differentiable functions. Consider a time discretization $0 = t_0 < t_1 < \cdots < t_m = T$. For $i = 0, 1, \ldots, m-1$,

$$X_{t_{i+1}} = X_{t_i} + \int_{t_i}^{t_{i+1}} b(X_t)\, dt + \int_{t_i}^{t_{i+1}} \sigma(X_t)\, dW_t.$$

The Euler scheme essentially approximates $b(X_t)$ and $\sigma(X_t)$ by $b(X_{t_i})$ and $\sigma(X_{t_i})$, respectively, for every $t \in [t_i, t_{i+1})$. To improve on the Euler scheme, we should examine the coefficients $b(X_t)$ and $\sigma(X_t)$ more carefully through Itô formula.

For a twice continuously differentiable function f, it follows from Itô formula that

$$f(X_t) = f(X_{t_i}) + \int_{t_i}^{t} \mathbb{L}^0 f(X_s)\, ds + \int_{t_i}^{t} \mathbb{L}^1 f(X_s)\, dW_s,$$

where

$$\mathbb{L}^0 f(x) = f'(x) b(x) + \frac{1}{2} f''(x) \sigma^2(x), \quad \mathbb{L}^1 f(x) = f'(x) \sigma(x).$$

Therefore,

$$X_{t_{i+1}} = X_{t_i} + b(X_{t_i})(t_{i+1} - t_i) + \sigma(X_{t_i})(W_{t_{i+1}} - W_{t_i}) + R,$$

where

$$
\begin{aligned}
R &= \int_{t_i}^{t_{i+1}} \left[\int_{t_i}^{t} \mathbb{L}^0 b(X_s)\, ds + \int_{t_i}^{t} \mathbb{L}^1 b(X_s)\, dW_s \right] dt \\
&\quad + \int_{t_i}^{t_{i+1}} \left[\int_{t_i}^{t} \mathbb{L}^0 \sigma(X_s)\, ds + \int_{t_i}^{t} \mathbb{L}^1 \sigma(X_s)\, dW_s \right] dW_t
\end{aligned}
$$

is the remainder term. Clearly, the Euler scheme ignores the remainder term R and keeps the two leading terms: (a) $\sigma(X_{t_i})(W_{t_{i+1}} - W_{t_i})$, which is of order $\sqrt{t_{i+1} - t_i}$; (b) $b(X_{t_i})(t_{i+1} - t_i)$, which is of order $t_{i+1} - t_i$.

Refinements of the Euler scheme retain some of the higher order terms in R in order to improve the approximation. Note that of all the terms in the remainder, the double stochastic integral is the leading term with order $t_{i+1} - t_i$ [other terms are of order $(t_{i+1} - t_i)^{3/2}$ or $(t_{i+1} - t_i)^2$]. If we approximate the double stochastic integral by

$$\mathbb{L}^1 \sigma(X_{t_i}) \int_{t_i}^{t_{i+1}} \int_{t_i}^{t} dW_s dW_t = \mathbb{L}^1 \sigma(X_{t_i}) \int_{t_i}^{t_{i+1}} (W_t - W_{t_i}) dW_t$$
$$= \frac{1}{2} \mathbb{L}^1 \sigma(X_{t_i})[(W_{t_{i+1}} - W_{t_i})^2 - (t_{i+1} - t_i)],$$

and add it to the Euler scheme while ignoring all other higher order terms, we obtain one of the *Milstein schemes*

$$\hat{X}_{t_{i+1}} = \hat{X}_{t_i} + b(\hat{X}_{t_i})(t_{i+1} - t_i) + \sigma(\hat{X}_{t_i}) \sqrt{t_{i+1} - t_i} Z_{i+1}$$
$$+ \frac{1}{2} \sigma'(\hat{X}_{t_i}) \sigma(\hat{X}_{t_i})(t_{i+1} - t_i)(Z_{i+1}^2 - 1), \qquad (9.3)$$

where $\{Z_1, \ldots, Z_m\}$ are iid standard normal random variables.

Remark 9.2. It is possible to define performance measures for numerical schemes [20]. Furthermore, it can be shown that the performance of the Milstein scheme is better than that of the Euler scheme, albeit not by much. It can also be shown that one can arrive at even better numerical schemes by retaining more terms from the remainder R.

9.4 The Lamperti Transform

The Milstein scheme (9.3) coincides with the plain Euler scheme if $\sigma'(x) = 0$, or equivalently, if $\sigma(x)$ is a constant function. This observation leads to the general understanding that the Euler scheme is more effective when the volatility is a constant. For a diffusion process with nonconstant volatility, the idea is to transform the process into a diffusion with constant volatility, apply the Euler scheme, and then revert back to the original process.

To be more concrete, consider a diffusion process X that satisfies the stochastic differential equation

$$dX_t = b(X_t)\, dt + \sigma(X_t)\, dW_t,$$

where σ is continuously differentiable and strictly positive. Define a function F and a process Y by

$$Y_t = F(X_t), \quad F(x) = \int \frac{1}{\sigma(x)}\, dx. \qquad (9.4)$$

Since $\sigma(x)$ is strictly positive, F is well defined and strictly increasing. In particular, F^{-1} exists and $X_t = F^{-1}(Y_t)$. The process Y is said to be the *Lamperti transform* of X. By the definition of F,

$$F'(x) = \frac{1}{\sigma(x)}, \quad F''(x) = -\frac{\sigma'(x)}{\sigma^2(x)}.$$

It follows from Itô formula that

$$dY_t = F'(X_t)\,dX_t + \frac{1}{2}F''(X_t)(dX_t)^2 = \left[\frac{b(X_t)}{\sigma(X_t)} - \frac{\sigma'(X_t)}{2}\right]dt + dW_t.$$

Substituting $F^{-1}(Y_t)$ for X_t, we arrive at $dY_t = a(Y_t)\,dt + dW_t$, where

$$a(y) = \frac{b(F^{-1}(y))}{\sigma(F^{-1}(y))} - \frac{\sigma'(F^{-1}(y))}{2}.$$

In other words, Y is a diffusion with constant volatility one. Below is the pseudocode for generating *one* discrete time sample path of $\{X_t\}$ through the Lamperti transform, given a time discretization $0 = t_0 < t_1 < \cdots < t_m = T$.

Pseudocode for Euler scheme through the Lamperti transform:

> set $Y_0 = F(X_0)$
> for $i = 0, 1, \ldots, m-1$
>   generate Z_{i+1} from $N(0,1)$
>   set $\hat{Y}_{t_{i+1}} = \hat{Y}_{t_i} + a(\hat{Y}_{t_i})(t_{i+1} - t_i) + \sqrt{t_{i+1} - t_i}\, Z_{i+1}$
>   set $\hat{X}_{t_{i+1}} = F^{-1}(\hat{Y}_{t_{i+1}})$

Example 9.1. Determine the Lamperti transform for the geometric Brownian motion

$$dX_t = rX_t\,dt + \sigma X_t\,dW_t$$

where r and σ are positive constants.

SOLUTION: In this case,

$$F(x) = \int \frac{1}{\sigma x}\,dx = \frac{1}{\sigma}\log x, \quad Y_t = F(X_t) = \frac{1}{\sigma}\log X_t,$$

and

$$X_t = F^{-1}(Y_t) = \exp\{\sigma Y_t\}.$$

By Itô formula,
$$dY_t = \left(\frac{r}{\sigma} - \frac{\sigma}{2}\right) dt + dW_t.$$

It is not difficult to check that the Euler scheme through the Lamperti transform leads to a scheme for X without discretization error. ∎

Example 9.2. Determine the Lamperti transform for the Cox–Ingersoll–Ross process
$$dX_t = a(b - X_t)\, dt + \sigma\sqrt{X_t}\, dW_t,$$

where a, b, σ are positive constants and $2ab \geq \sigma^2$.

SOLUTION: In this case,
$$F(x) = \int \frac{1}{\sigma\sqrt{x}}\, dx = \frac{2}{\sigma}\sqrt{x}, \quad Y_t = F(X_t) = \frac{2}{\sigma}\sqrt{X_t},$$

and
$$X_t = F^{-1}(Y_t) = \frac{\sigma^2}{4}Y_t^2.$$

By Itô formula,
$$
\begin{aligned}
dY_t &= \left(\frac{ab}{\sigma\sqrt{X_t}} - \frac{a\sqrt{X_t}}{\sigma} - \frac{\sigma}{4\sqrt{X_t}}\right) dt + dW_t \\
&= \left(\frac{4ab - \sigma^2}{2\sigma^2}\frac{1}{Y_t} - \frac{aY_t}{2}\right) dt + dW_t.
\end{aligned}
$$

Finally, we should remark that $2ab \geq \sigma^2$ is the necessary and sufficient condition for X_t to be strictly positive with probability one for all t; see for example, [18, Chapter 5.5.C]. If this condition fails, the Lamperti transform is no longer valid because the function $\sqrt{x}$ is not differentiable at $x = 0$. ∎

9.5 Numerical Examples

We present a few examples in this section to illustrate various aspects of numerical approximations to stochastic differential equations and explain how they can be combined with variance reduction techniques. Unless otherwise specified, W always stands for a standard Brownian motion.

Example 9.3. Comparison of Euler scheme and Milstein scheme. Suppose that X is a geometric Brownian motion with drift r and volatility σ. That is,

$$dX_t = rX_t\, dt + \sigma X_t\, dW_t.$$

Use the Euler scheme and Milstein scheme to approximate X_T. Compare the performance.

SOLUTION: Assume that the time discretization is $t_i = ih$ for $i = 0, 1, \ldots, m$ and $h = T/m$. For the Euler scheme, the approximation is given by

$$\hat{X}_0 = X_0, \quad \hat{X}_{t_{i+1}} = \hat{X}_{t_i} + r\hat{X}_{t_i}h + \sigma\hat{X}_{t_i}\sqrt{h}Z_{i+1},$$

while for the Milstien scheme,

$$\bar{X}_0 = X_0, \quad \bar{X}_{t_{i+1}} = \bar{X}_{t_i} + r\bar{X}_{t_i}h + \sigma\bar{X}_{t_i}\sqrt{h}Z_{i+1} + \frac{1}{2}\sigma^2\bar{X}_{t_i}h(Z_{i+1}^2 - 1),$$

where $\{Z_1, \ldots, Z_m\}$ are iid standard normal random variables. As the benchmark, we need a scheme without discretization error, which is simply

$$X_{t_{i+1}} = X_{t_i}\exp\left\{\left(r - \frac{1}{2}\sigma^2\right)h + \sigma\sqrt{h}Z_{i+1}\right\}.$$

For comparison, we estimate the errors $E|\hat{X}_T - X_T|$ and $E|\bar{X}_T - X_T|$. Such quantities are often used in the literature as a performance measure. The numerical results are reported in Table 9.1 with parameters

$$r = 0.12, \ \sigma = 0.2, \ T = 1, \ X_0 = 50.$$

For each value of m, we use a large sample size of $n = 10,000,000$ to get an accurate estimate of the error.

Table 9.1: Numerical approximation: Euler versus Milstein

	$m = 5$	$m = 10$	$m = 20$	$m = 40$	$m = 60$	$m = 80$	$m = 100$
Euler	0.5615	0.3994	0.2832	0.2007	0.1640	0.1421	0.1271
Milstein	0.2208	0.1121	0.0565	0.0284	0.0189	0.0142	0.0114

From Figure 9.1, it is clear that the error from the Milstein scheme is of order h and the error from the Euler scheme is of order $\sqrt{h}$ (the slope of the line in the log-log scale is roughly $1/2$). Those rates of convergence can be proved in much greater generality [20]. ∎

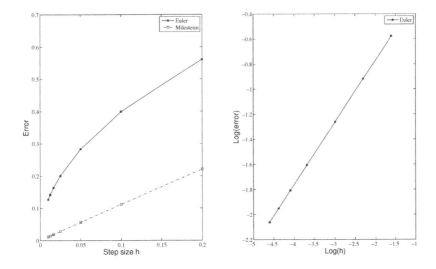

Figure 9.1: Numerical approximation: Euler versus Milstein.

Example 9.4. No discretization error. Consider a discretely monitored lookback call option with maturity T and floating strike price. The option payoff is

$$X = S_T - \min_{i=1,\ldots,m} S_{t_i}$$

where $0 = t_0 < t_1 < \cdots < t_m = T$ are given dates. Assume that the risk-free interest rate r is a constant and the stock price S satisfies the stochastic differential equation

$$dS_t = rS_t \, dt + \theta(t) S_t \, dW_t$$

for some deterministic function $\theta(t)$ under the risk-neutral probability measure. Design a discretization scheme without discretization error and estimate the option price.

SOLUTION: Define $Y_t = \log S_t$. Then $Y_0 = \log S_0$. It follows from Itô formula that

$$dY_t = \frac{1}{S_t} dS_t - \frac{1}{2S_t^2}(dS_t)^2 = \left(r - \frac{1}{2}\theta^2(t)\right) dt + \theta(t)\, dW_t.$$

Therefore,

$$
\begin{aligned}
Y_{t_{i+1}} &= Y_{t_i} + \int_{t_i}^{t_{i+1}} \left(r - \frac{1}{2}\theta^2(t) \right) dt + \int_{t_i}^{t_{i+1}} \theta(t)\, dW_t \\
&= Y_{t_i} + \int_{t_i}^{t_{i+1}} \left(r - \frac{1}{2}\theta^2(t) \right) dt + \sqrt{\int_{t_i}^{t_{i+1}} \theta^2(t)\, dt} \cdot Z_{i+1} \\
&= Y_{t_i} + r(t_{i+1} - t_i) - \frac{1}{2}\sigma_i^2 + \sigma_i Z_{i+1},
\end{aligned}
$$

where

$$
\sigma_i^2 = \int_{t_i}^{t_{i+1}} \theta^2(t)\, dt,
$$

and $\{Z_1, \ldots, Z_m\}$ are iid standard normal random variables. Letting $S_{t_i} = \exp\{Y_{t_i}\}$, we obtain a discretization scheme for S without discretization error. Below is the pseudocode, in which Y_i and S_i stand for Y_{t_i} and S_{t_i}, respectively.

Pseudocode for lookback call option without discretization error:

> for $k = 1, 2, \ldots, n$
>> set $Y_0 = \log S_0$
>> for $i = 0, 1, \ldots, m-1$
>>> generate Z_{i+1} from $N(0,1)$
>>> set $Y_{i+1} = Y_i + r(t_{i+1} - t_i) - \sigma_i^2/2 + \sigma_i Z_{i+1}$
>>> set $S_{i+1} = \exp\{Y_{i+1}\}$
>>> set $H_k = e^{-rT}[S_m - \min(S_1, \ldots, S_m)]$
>> compute the estimate $\hat{v} = \dfrac{1}{n} \sum_{i=1}^{n} H_i$
>> compute the standard error S.E. $= \sqrt{\dfrac{1}{n(n-1)} \left(\sum_{i=1}^{n} H_i^2 - n\hat{v}^2 \right)}.$

For numerical simulation, we let $\theta(t) = \sqrt{a^2 + b^2 \sin(2\pi t/T)}$ for some positive constants a and b such that $a > b$. In this case,

$$
\sigma_i^2 = a^2(t_{i+1} - t_i) + \frac{b^2 T}{2\pi}[\cos(2\pi t_i/T) - \cos(2\pi t_{i+1}/T)].
$$

The simulation results are reported in Table 9.2 for

$$
S_0 = 20, \ r = 0.05, \ T = 1, \ a = 0.4, \ b = 0.3.
$$

We compare with the corresponding plain Euler scheme for $m = 5, 10, 20$, respectively (for each m, the time discretization in the Euler scheme is defined to be $0 = t_0 < t_1 < \cdots < t_m = T$). The sample size n is set to be $n = 1,000,000$ for accurate comparison.

Table 9.2: Scheme without discretization error versus Euler scheme

	$m = 5$		$m = 10$		$m = 20$	
	Euler	Exact	Euler	Exact	Euler	Exact
Estimate	4.2801	4.1409	4.9888	4.9529	5.4373	5.4239
S.E.	0.0055	0.0053	0.0059	0.0061	0.0064	0.0062

The standard errors associated with both schemes are nearly identical. However, when m is small the estimates differ because of the large bias of the Euler scheme. It is not surprising that as m increases, the bias of the Euler scheme diminishes and the two schemes become indistinguishable.

If one were to consider the discretely monitored lookback option as an approximation to the continuous time lookback call option with maturity T and payoff

$$S_T - \min_{0 \le t \le T} S_t,$$

then the scheme would be an example where the underlying process is simulated without discretization error but the estimate is still biased because the payoff function is only approximated. ∎

Example 9.5. Simulating a process with sign constraints. The majority of the diffusion processes that are used for modeling stock prices, interest rate, volatility, and other financial entities are nonnegative. This property may fail for the time discretized approximating processes, which sometimes results in necessary modifications of the discretization scheme. To be more concrete, consider the Cox–Ingersoll–Ross interest rate model [6] where the short rate $\{r_t\}$ satisfies the stochastic differential equation

$$dr_t = a(b - r_t)\, dt + \sigma \sqrt{r_t}\, dW_t$$

for some positive constants a, b, and σ such that $2ab \ge \sigma^2$. Estimate the price of a zero-coupon bond with maturity T and payoff \$1.

SOLUTION: The price of the zero-coupon bond is given by the pricing formula (8.7) with $X = 1$:

$$v = E\left[\exp\left\{-\int_0^T r_t\, dt\right\}\right]. \tag{9.5}$$

The plain Euler scheme for $\{r_t\}$ is as follows. Let $\hat{r}_0 = r_0$ and define recursively

$$\hat{r}_{t_{i+1}} = \hat{r}_{t_i} + a(b - \hat{r}_{t_i})(t_{i+1} - t_i) + \sigma\sqrt{\hat{r}_{t_i}(t_{i+1} - t_i)}\, Z_{i+1}$$

for $i = 0, 1, \ldots, m - 1$. Approximating the integral in (9.5) by the corresponding Riemann sum, the estimate for the bond price is just the sample average of iid copies of

$$H = \exp\left\{ -\sum_{i=0}^{m-1} \hat{r}_{t_i}(t_{i+1} - t_i) \right\}.$$

This scheme does not work universally since there is a small chance that $\hat{r}_{t_i}$ might be negative and its square root is not well defined.

There are many ways to modify the plain Euler scheme, and most of them do not make much difference in practice. We consider three legitimate modifications.

$$\hat{r}_{t_{i+1}} = \max\{0, \hat{r}_{t_i} + a(b - \hat{r}_{t_i})(t_{i+1} - t_i) + \sigma\sqrt{\hat{r}_{t_i}(t_{i+1} - t_i)}\, Z_{i+1}; \quad \text{(A)}$$

$$\hat{r}_{t_{i+1}} = \hat{r}_{t_i} + a(b - \hat{r}_{t_i})(t_{i+1} - t_i) + \sigma\sqrt{\max\{\hat{r}_{t_i}, 0\}(t_{i+1} - t_i)}\, Z_{i+1}; \quad \text{(B)}$$

$$\hat{r}_{t_{i+1}} = \hat{r}_{t_i} + a(b - \hat{r}_{t_i})(t_{i+1} - t_i) + \sigma\sqrt{|\hat{r}_{t_i}|(t_{i+1} - t_i)}\, Z_{i+1}. \quad \text{(C)}$$

The numerical results are presented in Table 9.3. We test the schemes for

$$a = 1, \quad b = 0.05, \quad m = 20, \quad r_0 = 0.08, \quad T = 1, \quad t_i = iT/m,$$

and a range of σ's. All estimates are based on $n = 1,000,000$ samples with common sequences of $\{Z_i\}$.

Table 9.3: Modifications of Euler scheme

	$\sigma = 0.1$	$\sigma = 0.15$	$\sigma = 0.2$	$\sigma = 0.25$	$\sigma = 0.3$
Scheme (A)	0.9332	0.9332	0.9333	0.9334	0.9335
Scheme (B)	0.9332	0.9332	0.9333	0.9334	0.9336
Scheme (C)	0.9332	0.9332	0.9333	0.9334	0.9336
Fraction	0.00%	0.00%	0.17%	2.11%	8.01%

The standard errors for these estimates, which are not reported, all range between 1.0×10^{-5} and 3.0×10^{-5}. The entry "fraction" records the fraction of the time discretized sample paths that would have reached negative if the plain Euler scheme had been used. ∎

Example 9.6. Method of conditioning. Suppose that under the risk-neutral probability measure, the stock price satisfies the stochastic differential equation

$$\frac{dS_t}{S_t} = r_t\, dt + \sigma\, dW_t.$$

The short rate $\{r_t\}$ is assumed to be mean-reverting and satisfy the stochastic differential equation

$$dr_t = a(b - r_t)\, dt + \theta r_t dB_t,$$

for some positive constants a, b, and θ. Assume that (W, B) is a two-dimensional Brownian motion with covariance matrix

$$\Sigma = \begin{bmatrix} 1 & \rho \\ \rho & 1 \end{bmatrix}.$$

Estimate the price of a call option with maturity T and strike price K.

SOLUTION: It is not difficult to verify via Itô formula that the terminal stock price is (see also Exercise 8.13)

$$
\begin{aligned}
S_T &= S_0 \exp\left\{ \int_0^T r_t\, dt - \frac{1}{2}\sigma^2 T + \sigma W_T \right\} \\
&= S_0 \exp\left\{ \left(\bar{r} - \frac{1}{2}\sigma^2\right) T + \sigma W_T \right\},
\end{aligned}
$$

where

$$\bar{r} = \frac{1}{T}\int_0^T r_t\, dt$$

is the average interest rate (which a random variable). The price of the call option is $v = E\left[e^{-\bar{r}T}(S_T - K)^+\right]$, where the expected value is taken under the risk-neutral probability measure.

The Euler scheme of the two-dimensional process (S_t, r_t) is straightforward. Let $0 = t_0 < t_1 < \cdots < t_m = T$ be a time discretization. Let $\hat{S}_0 = S_0$ and $\hat{r}_0 = r_0$. Then for $i = 0, 1, \ldots, m - 1$

$$
\begin{aligned}
\hat{S}_{t_{i+1}} &= \hat{S}_{t_i} + \hat{r}_{t_i}\hat{S}_{t_i}(t_{i+1} - t_i) + \sigma\hat{S}_{t_i}\sqrt{t_{i+1} - t_i}\,Z_{i+1}, \\
\hat{r}_{t_{i+1}} &= \hat{r}_{t_i} + a(b - \hat{r}_{t_i})(t_{i+1} - t_i) + \theta\hat{r}_{t_i}\sqrt{t_{i+1} - t_i}\,Y_{i+1},
\end{aligned}
$$

where (Z_i, Y_i) are iid jointly normal random vectors with mean 0 and covariance matrix Σ. For each sample path, $\bar{r}$ can be approximated by

$$\bar{R} = \frac{1}{T}\sum_{i=0}^{m-1} \hat{r}_{t_i}(t_{i+1} - t_i).$$

The corresponding plain Monte Carlo estimate is just the sample average of iid copies of $e^{-\bar{R}T}(\hat{S}_T - K)^+$.

A more efficient algorithm is to combine the Euler scheme with the method of conditioning. The key observation is that the conditional expectation

$$h(\bar{r}, B_T) = E[e^{-\bar{r}T}(S_T - K)^+ | \bar{r}, B_T]$$

is explicitly solvable. Indeed, define

$$Q_t = \frac{1}{\sqrt{1 - \rho^2}}(W_t - \rho B_t).$$

It is not difficult to show that (Q, B) is a two-dimensional standard Brownian motion; see Exercise 2.16. In other words, Q is a standard Brownian motion and is independent of B. Substituting $\rho B_t + \sqrt{1 - \rho^2}Q_t$ for W_t, we have

$$
\begin{aligned}
S_T &= S_0 \exp\left\{\left(\bar{r} - \frac{1}{2}\sigma^2\right)T + \sigma\rho B_T + \sigma\sqrt{1 - \rho^2}\,Q_T\right\} \\
&= X_0 \exp\left\{\left(\bar{r} - \frac{1}{2}\sigma^2(1 - \rho^2)\right)T + \sigma\sqrt{1 - \rho^2}\,Q_T\right\},
\end{aligned}
$$

where

$$X_0 = S_0 \exp\left\{\rho\sigma B_T - \frac{1}{2}\rho^2\sigma^2 T\right\}.$$

Therefore, given $\bar{r}$ and B_T, S_T is the terminal value of a geometric Brownian motion with initial value X_0, drift $\bar{r}$, and volatility $\sigma\sqrt{1 - \rho^2}$. Consequently, $h(\bar{r}, B_T)$ is the corresponding price of a call option with strike price K and maturity T. That is,

$$h(\bar{r}, B_T) = \text{BLS\_Call}(X_0, K, T, \bar{r}, \sigma\sqrt{1 - \rho^2}), \tag{9.6}$$

where BLS_Call denotes the price of a call option under the classical Black–Scholes model; see Example 2.1. By the tower property, the option price is

$$v = E[h(\bar{r}, B_T)].$$

This leads to a very simple algorithm in which the stock price S does not need to be simulated at all. Only the process $\{r_t\}$ will be approximated via (say) the Euler scheme. Sampling B_T does not incur extra computational cost—it is just a by-product of the Euler scheme for $\{r_t\}$. Below is the pseudocode. Abusing notation, we use $\hat{r}_i$ to denote $\hat{r}_{t_i}$.

Pseudocode for Euler scheme and method of conditioning:

for $k = 1, 2, \ldots, n$

    set $\hat{r}_0 = r_0$ and $L = 0$

    for $i = 0, 1, \ldots, m - 1$

        generate sample Y_{i+1} from $N(0,1)$

        set $\hat{r}_{i+1} = \hat{r}_i + a(b - \hat{r}_i)(t_{i+1} - t_i) + \theta \hat{r}_i \sqrt{t_{i+1} - t_i}\, Y_{i+1}$

        set $L = L + \sqrt{t_{i+1} - t_i}\, Y_{i+1}$

    set $B_T = L$

    set $R = [(t_1 - t_0)\hat{r}_0 + \cdots + (t_m - t_{m-1})\hat{r}_{m-1}]/T$

    set $X_0 = S_0 \exp\{\rho\sigma B_T - \rho^2 \sigma^2 T/2\}$

    set $H_k = C(X_0, K, T, R, \sigma\sqrt{1 - \rho^2})$

compute the estimate $\hat{v} = \dfrac{1}{n}\displaystyle\sum_{i=1}^{n} H_i$

compute the standard error S.E. $= \sqrt{\dfrac{1}{n(n-1)}\left(\displaystyle\sum_{i=1}^{n} H_i^2 - n\hat{v}^2\right)}.$

The numerical results are reported in Table 9.4 for both the plain Monte Carlo scheme and the hybrid algorithm combining the Euler scheme and the method of conditioning. The parameters are given by

$$S_0 = 50,\ r_0 = 0.12,\ b = 0.10,\ a = 2,\ \theta = \sigma = 0.2,\ T = 1,\ \rho = 0.3.$$

The time interval is divided into $m = 50$ subintervals of equal length and the sample size is $n = 10000$.

Table 9.4: Method of conditioning with stochastic short rate

	$K = 45$		$K = 50$		$K = 55$	
	Plain	Hybrid	Plain	Hybrid	Plain	Hybrid
Estimate	10.3599	10.2925	6.8799	6.8584	4.3659	4.3207
S.E.	0.0925	0.0285	0.0822	0.0240	0.0685	0.0188

Clearly, the method of conditioning is more efficient than the plain Monte Carlo scheme when it comes to variance reduction. It is also possible to replace the Euler scheme for $\{r_t\}$ by the Milstein scheme to reduce the bias. But given the relatively small sample size, the performance is indistinguishable from the Euler scheme's. ∎

Example 9.7. Control variate method by means of artificial dynamics.
Consider a stochastic volatility model (the volatility is denoted by θ) where
the risk-free interest rate is a constant r and under the risk-neutral probability measure,

$$
\begin{aligned}
dS_t &= rS_t\, dt + \theta_t S_t\, dW_t, \\
d\theta_t &= a(\Theta - \theta_t)\, dt + \beta\, dB_t,
\end{aligned}
$$

for some positive constants a, Θ, and β. Here (W, B) is a two-dimensional
Brownian motion with covariance matrix

$$
\Sigma = \begin{bmatrix} 1 & \rho \\ \rho & 1 \end{bmatrix}.
$$

Estimate the price of a call option with strike price K and maturity T.

SOLUTION: By Itô formula, $Y_t = \log S_t$ satisfies the stochastic differential
equation

$$
dY_t = \left(r - \frac{1}{2}\theta_t^2 \right) dt + \theta_t\, dW_t.
$$

A very simple scheme for estimating the call option price is to apply the
Euler scheme on the two-dimensional process (Y, θ). More precisely, let $0 = t_0 < t_1 < \cdots < t_m = T$ be a time discretization. Define $\hat{Y}_0 = Y_0 = \log S_0$
and $\hat{\theta}_0 = \theta_0$. For $i = 0, 1, \ldots, m-1$, recursively define

$$
\begin{aligned}
\hat{Y}_{t_{i+1}} &= \hat{Y}_{t_i} + \left(r - \frac{1}{2}\hat{\theta}_{t_i}^2 \right)(t_{i+1} - t_i) + \hat{\theta}_{t_i}\sqrt{t_{i+1} - t_i}\, Z_{i+1}, \\
\hat{\theta}_{t_{i+1}} &= \hat{\theta}_{t_i} + a(\Theta - \hat{\theta}_{t_i})(t_{i+1} - t_i) + \beta\sqrt{t_{i+1} - t_i}\, R_{i+1},
\end{aligned}
$$

where $\{(Z_1, R_1), \ldots, (Z_m, R_m)\}$ are iid jointly normal random vectors with
distribution $N(0, \Sigma)$. A plain Monte Carlo estimate for the call option price
is just the sample average of iid copies of $X = e^{-rT}(\exp\{\hat{Y}_T\} - K)^+$.

To improve the efficiency, one can use the control variate method by
introducing an *artificial* stochastic process $\bar{Y}$ that is recursively defined by
$\bar{Y}_0 = Y_0 = \log S_0$ and

$$
\bar{Y}_{t_{i+1}} = \bar{Y}_{t_i} + \left(r - \frac{1}{2}\sigma^2 \right)(t_{i+1} - t_i) + \sigma\sqrt{t_{i+1} - t_i}\, Z_{i+1}
$$

for $i = 0, 1, \ldots, m-1$. This process $\bar{Y}$ uses the *same* sequence $\{Z_1, \ldots, Z_m\}$,
and some *constant* volatility σ. One can think of $\bar{Y}$ as the logarithm of the

price of a virtual stock that follows a classical geometric Brownian motion with drift r and volatility σ. Then the discounted payoff of the call option for this virtual stock, namely,

$$V = e^{-rT}(\exp(\bar{Y}_T) - K)^+$$

can be used as a control variate, whose expected value is just the classical Black–Scholes call option price BLS_Call(S_0, K, T, r, σ). The control variate estimate for the call option price is the sample average of iid copies of

$$X - b[V - \text{BLS\_Call}(S_0, K, T, r, \sigma)]$$

for some constant b. We opt to use $b = \hat{b}^*$, which is the sample estimate of the optimal coefficient; see formula (6.2).

The value of σ can be arbitrary in theory. However, it is usually chosen to be a typical value of θ in order to achieve a higher level of variance reduction. In our example, the process θ is a mean reverting Ornstein–Uhlenbeck process. Therefore, it is reasonable to let $\sigma = \Theta$, the long run average of θ. Below is the pseudocode. As before, we use $\hat{Y}_i$ and $\hat{\theta}_i$ to denote $\hat{Y}_{t_i}$ and $\hat{\theta}_{t_i}$, respectively.

Pseudocode for control variate method:

set $\sigma = \Theta$

for $k = 1, 2, \ldots, n$

    set $\hat{Y}_0 = \log S_0$, $\bar{Y}_0 = \log S_0$, and $\hat{\theta}_0 = \theta_0$

    for $i = 0, 1, \ldots, m - 1$

        generate iid sample Z_{i+1} and U_{i+1} from $N(0, 1)$

        set $R_{i+1} = \rho Z_{i+1} + \sqrt{1 - \rho^2} U_{i+1}$

        set $\hat{Y}_{i+1} = \hat{Y}_i + (r - \hat{\theta}_i^2/2)(t_{i+1} - t_i) + \hat{\theta}_i \sqrt{t_{i+1} - t_i} Z_{i+1}$

        set $\bar{Y}_{i+1} = \bar{Y}_i + (r - \sigma^2/2)(t_{i+1} - t_i) + \sigma \sqrt{t_{i+1} - t_i} Z_{i+1}$

        set $\hat{\theta}_{i+1} = \hat{\theta}_i + a(\Theta - \hat{\theta}_i)(t_{i+1} - t_i) + \beta \sqrt{t_{i+1} - t_i} R_{i+1}$

    set $X_k = e^{-rT}(\exp\{\hat{Y}_m\} - K)^+$;

    set $Q_k = e^{-rT}(\exp\{\bar{Y}_m\} - K)^+ - \text{BLS\_Call}(S_0, K, T, r, \sigma)$

compute $\hat{b}^*$ from formula (6.2) [with Y replaced by Q]

for $k = 1, 2, \ldots, n$

    set $H_k = X_k - \hat{b}^* Q_k$

compute the estimate $\hat{v} = \dfrac{1}{n} \sum_{i=1}^{n} H_i$

compute the standard error S.E. $= \sqrt{\dfrac{1}{n(n-1)} \left(\sum_{i=1}^{n} H_i^2 - n\hat{v}^2 \right)}$.

We compare the plain Euler scheme with the control variate method. The numerical results are reported in table 9.5 for

$$r = 0.05,\ S_0 = 50,\ K = 50,\ a = 3,\ \beta = 0.1,\ \theta_0 = 0.25,\ T = 1,$$

$$\Theta = 0.2,\ \rho = 0.5,\ m = 50,\ t_i = iT/m,\ n = 10000.$$

Table 9.5: Control variate method for a stochastic volatility model

	Euler	Control Variate				
		$\sigma = \Theta/5$	$\sigma = \Theta/2$	$\sigma = \Theta$	$\sigma = 2\Theta$	$\sigma = 5\Theta$
Estimate	5.5301	5.5044	5.5621	5.5688	5.5430	5.6202
S.E.	0.0867	0.0402	0.0252	0.0181	0.0171	0.0413

We have tested a wide range of σ to investigate its effect on variance reduction. The choice of $\sigma = \Theta$ has turned out to be nearly optimal. ∎

Example 9.8. Importance sampling and cross-entropy method. Assume that the risk-free interest rate r is a constant and the underlying stock price is a constant elasticity of variance (CEV) process under the risk-neutral probability measure, that is,

$$dS_t = rS_t\, dt + \sigma S_t^{\gamma}\, dW_t$$

for some $0.5 \le \gamma < 1$. Estimate the price of a call option with maturity T and strike price K.

SOLUTION: Let $X_t = e^{-rt} S_t$. Then $X_0 = S_0$ and it follows from Itô formula that

$$dX_t = \sigma e^{-r(1-\gamma)t} X_t^{\gamma}\, dW_t.$$

The price of the call option is

$$v = E[e^{-rT}(S_T - K)^+] = E[(X_T - e^{-rT}K)^+].$$

In order to guarantee that the time discretized approximating process $\hat{X}$ is nonnegative, we modify the plain Euler scheme slightly. Let $0 = t_0 < t_1 < \cdots < t_m = T$ be a time discretization. Given $\hat{X}_0 = S_0$, define recursively

$$\hat{X}_{t_{i+1}} = \max\{0, \hat{X}_{t_i} + \sigma e^{-r(1-\gamma)t_i}\, \hat{X}_{t_i}^{\gamma} \sqrt{\Delta t_{i+1}} \cdot Z_{i+1}\}$$

for $i = 0, 1, \ldots, m-1$, where $\{Z_1, \ldots, Z_m\}$ are iid standard normal random variables and $\Delta t_{i+1} = t_{i+1} - t_i$. Note that if $\hat{X}$ ever reaches zero, it will stay at zero. This is exactly what we want because zero is an absorbing state for the original process X (loosely speaking, when X reaches zero, the volatility becomes zero and X stays at zero). The plain Monte Carlo estimate is the sample average of iid copies of

$$(\hat{X}_T - e^{-rT}K)^+,$$

which is indeed a function of $Y = (Z_1, \ldots, Z_m)$, say $h(Y)$. Note that Y is a jointly normal random vector with distribution $N(0, I_m)$.

Assuming that the alternative sampling distribution is $N(\theta, I_m)$ where $\theta = (\theta_1, \ldots, \theta_m) \in \mathbb{R}^m$ and the strike price K is not overly large, the basic cross-entropy method can be adopted to determine a nearly optimal tilting parameter $\hat{\theta}$. More precisely, by Lemma 7.1

$$\hat{\theta} = \frac{\sum_{k=1}^N h(Y_k) Y_k}{\sum_{k=1}^N h(Y_k)},$$

where Y_k's are iid jointly normal random vectors with distribution $N(0, I_m)$. Below is the pseudocode. As usual, we use $\hat{X}_i$ to denote $\hat{X}_{t_i}$.

Pseudocode for call price by the basic cross-entropy method:

generate iid pilot samples $Y_1, \ldots, Y_N$ from $N(0, I_m)$
set $\hat{\theta} = (\hat{\theta}_1, \ldots, \hat{\theta}_m) = \sum_{k=1}^N h(Y_k) Y_k / \sum_{k=1}^N h(Y_k)$
for $k = 1, 2, \ldots, n$
    for $i = 0, 1, \ldots, m-1$
        generate Z_{i+1} from $N(\hat{\theta}_{i+1}, 1)$
        set $\hat{X}_{i+1} = \max\{0, \hat{X}_i + \sigma e^{-r(1-\gamma)t_i} \hat{X}_i^\gamma \sqrt{\Delta t_{i+1}} \cdot Z_{i+1}\}$
    compute the discounted payoff multiplied by the likelihood ratio

$$H_k = (\hat{X}_m - e^{-rT}K)^+ \cdot \exp\left\{ -\sum_{i=1}^m \hat{\theta}_i Z_i + \frac{1}{2} \sum_{i=1}^m \hat{\theta}_i^2 \right\}$$

compute the estimate $\hat{v} = \dfrac{1}{n} \sum_{i=1}^n H_i$

compute the standard error S.E. $= \sqrt{\dfrac{1}{n(n-1)} \left(\sum_{i=1}^n H_i^2 - n\hat{v}^2 \right)}.$

We compare the plain Monte Carlo scheme with the basic cross-entropy method. The numerical results are reported in Table 9.6 for

$$S_0 = 50, \ r = 0.05, \ \sigma = 0.2, \ T = 1, \ m = 50, \ n = 10000, \ N = 2000.$$

The basic cross-entropy scheme is clearly more efficient.

Table 9.6: Basic cross-entropy method for a CEV model

$K = 50$	$\gamma = 0.5$		$\gamma = 0.7$		$\gamma = 0.9$	
	Plain	IS	Plain	IS	Plain	IS
Estimate	2.4455	2.4603	2.7788	2.7716	3.9644	4.0485
S.E.	0.0135	0.0065	0.0257	0.0127	0.0503	0.0214
$K = 55$	$\gamma = 0.5$		$\gamma = 0.7$		$\gamma = 0.9$	
	Plain	IS	Plain	IS	Plain	IS
Estimate	0.0272	0.0298	0.4131	0.4075	1.7319	1.7270
S.E.	0.0017	0.0003	0.0106	0.0035	0.0346	0.0127

Both the basic cross-entropy method and the plain Monte Carlo scheme will fail as the strike price increases to a certain level. A remedy is the general iterative cross-entropy method. However, a decent initial tilting parameter $\hat{\theta}^0$ is required for this to work. Even though it is straightforward to use the general initialization technique outlined in Section 7.2.3 to produce such an initial tilting parameter, we should present an alternative approach that does not require extra simulation. Assume that

$$\hat{\theta}^0 = x(\sqrt{\Delta t_1}, \sqrt{\Delta t_2}, \dots, \sqrt{\Delta t_m})$$

for some constant x. Under the alternative sampling distribution $N(\hat{\theta}^0, I_m)$, the discretization scheme becomes

$$\hat{X}_{t_{i+1}} = \max\{0, \hat{X}_{t_i} + \sigma e^{-r(1-\gamma)t_i} \hat{X}_{t_i}^{\gamma} \sqrt{\Delta t_{i+1}} \cdot \bar{Z}_{i+1}\},$$

where $\bar{Y} = (\bar{Z}_1, \dots, \bar{Z}_m)$ is a jointly normal random vector with distribution $N(\hat{\theta}^0, I_m)$. What we want is to choose x so that $E[\hat{X}_T]$ is approximately $e^{-rT}K$, and thus a reasonable fraction of samples will yield strictly positive payoffs. To this end, rewrite $\bar{Z}_{i+1} = x\sqrt{\Delta t_{i+1}} + R_{i+1}$ for each i, where $\{R_1, \dots, R_m\}$ are iid standard normals. We arrive at

$$\hat{X}_{t_{i+1}} = \max\{0, \hat{X}_{t_i} + \sigma e^{-r(1-\gamma)t_i} \hat{X}_{t_i}^{\gamma} (x\Delta t_{i+1} + \sqrt{\Delta t_{i+1}} \cdot R_{i+1})\}.$$

It is easy to see that $\hat{X}$ is the time discretized approximation of $\bar{X}$ where

$$d\bar{X}_t = x\sigma e^{-r(1-\gamma)t} \bar{X}_t^{\gamma} \, dt + \sigma e^{-r(1-\gamma)t} \bar{X}_t^{\gamma} \, dW_t, \quad \bar{X}_0 = X_0 = S_0.$$

Therefore, it suffices to choose x so that $E[\bar{X}_T]$ is approximately $e^{-rT}K$. Even though the expected value of $E[\bar{X}_T]$ is not explicitly known, we observe that by the martingale property of stochastic integrals

$$
\begin{aligned}
E[\bar{X}_t] &= \bar{X}_0 + x\sigma \int_0^t e^{-r(1-\gamma)s} E[\bar{X}_s^\gamma]\, ds \\
&\approx S_0 + x\sigma \int_0^t e^{-r(1-\gamma)s} (E[\bar{X}_s])^\gamma\, ds.
\end{aligned}
\tag{9.7}
$$

Here we have approximated $E[\bar{X}_s^\gamma]$ by $(E[\bar{X}_s])^\gamma$. These two quantities would be the same if $\gamma = 1$ or $\bar{X}_s$ were a constant. This implies that the approximation is good when γ is close to 1. When γ is away from 1, the variance of $\bar{X}_s$ is often relatively small, which again justifies the validity of the approximation. Now (9.7) implies that $E[\bar{X}_t]$ is approximately $f(t)$, where $f(t)$ satisfies

$$
f(t) = S_0 + x\sigma \int_0^t e^{-r(1-\gamma)s} f^\gamma(s)\, ds,
$$

or equivalently,

$$
\frac{1}{f^\gamma}\frac{df}{dt} = x\sigma e^{-r(1-\gamma)t}, \quad f(0) = S_0.
$$

Solving this ordinary differential equation explicitly, we obtain

$$
f^{1-\gamma}(t) - S_0^{1-\gamma} = \frac{x\sigma}{r}\left[1 - e^{-r(1-\gamma)t}\right].
$$

Letting $f(T) = e^{-rT}K$, we arrive at

$$
x = \frac{r}{\sigma}\frac{(e^{-rT}K)^{1-\gamma} - S_0^{1-\gamma}}{1 - e^{-r(1-\gamma)T}}.
\tag{9.8}
$$

Below is the pseudocode.

Pseudocode for the general iterative cross-entropy method:

set x by formula (9.8)

initialize $\hat{\theta}^0 = x(\sqrt{\Delta t_1}, \ldots, \sqrt{\Delta t_m})$ and set the iteration counter $j = 0$

(*) generate N iid samples $\bar{Y}_1, \ldots, \bar{Y}_N$ from $N(\hat{\theta}^j, I_m)$

set $\hat{\theta}^{j+1} = \sum_{k=1}^N h(\bar{Y}_k)e^{-\langle \hat{\theta}^j, \bar{Y}_k \rangle}\bar{Y}_k / \sum_{k=1}^N h(\bar{Y}_k)e^{-\langle \hat{\theta}^j, \bar{Y}_k \rangle}$

set the iteration counter $j = j + 1$

if $j = $ IT_NUM set $\hat{\theta} = \hat{\theta}^j$ and continue, otherwise go to step (*)

for $k = 1, 2, \ldots, n$

    set $\hat{X}_0 = S_0$

    for $i = 0, 1, \ldots, m - 1$

        generate Z_{i+1} from $N(\hat{\theta}_{i+1}, 1)$

        set $\hat{X}_{i+1} = \max\{0, \hat{X}_i + \sigma e^{-r(1-\gamma)t_i} \hat{X}_i^\gamma \sqrt{\Delta t_{i+1}} \cdot Z_{i+1}\}$

    compute the discounted payoff multiplied by the likelihood ratio

$$H_k = (\hat{X}_m - e^{-rT} K)^+ \cdot \exp\left\{ -\sum_{i=1}^m \hat{\theta}_i Z_i + \frac{1}{2} \sum_{i=1}^m \hat{\theta}_i^2 \right\}$$

compute the estimate $\hat{v} = \dfrac{1}{n} \sum_{i=1}^n H_i$

compute the standard error S.E. $= \sqrt{\dfrac{1}{n(n-1)} \left(\sum_{i=1}^n H_i^2 - n\hat{v}^2 \right)}.$

Table 9.7: Iterative cross-entropy method for a CEV model

$K = 57$	$\gamma = 0.5$		$\gamma = 0.7$		$\gamma = 0.9$	
	Plain	IS	Plain	IS	Plain	IS
Estimate	0.0005	6.2068×10^{-4}	0.1416	0.1315	1.1537	1.1931
S.E.	0.0002	7.7895×10^{-6}	0.0061	0.0012	0.0282	0.0084
$K = 60$	$\gamma = 0.5$		$\gamma = 0.7$		$\gamma = 0.9$	
	Plain	IS	Plain	IS	Plain	IS
Estimate	0	1.3247×10^{-7}	0.0171	0.0157	0.6100	0.6072
S.E.	0	2.1961×10^{-9}	0.0018	0.0002	0.0202	0.0048

We present the numerical results in Table 9.7 with the same parameters, except that the strike price is larger. The sample size is $n = 10000$ and the pilot sample size is $N = 2000$ with IT_NUM $= 5$ iterations. ∎

Example 9.9. Importance sampling and cross-entropy method. Consider the Heston model where the risk-free interest rate is a constant r and under the risk-neutral probability measure

$$\begin{aligned}
dS_t &= rS_t \, dt + \sqrt{\theta_t} S_t \, dW_t, \\
d\theta_t &= a(b - \theta_t) \, dt + \sigma \sqrt{\theta_t} \, dB_t,
\end{aligned}$$

for some positive constants a, b, and σ such that $2ab \geq \sigma^2$. Here (W, B) is a two-dimensional Brownian motion with covariance matrix

$$\Sigma = \begin{bmatrix} 1 & \rho \\ \rho & 1 \end{bmatrix}.$$

Estimate the price of a call option with maturity T and strike price K.

SOLUTION: The call option price is $v = E[e^{-rT}(S_T - K)^+]$. A slightly modified Euler scheme for the two-dimensional process (S, θ), which ensures that $\hat\theta$ stays nonnegative, is straightforward. Given a time discretization $0 = t_0 < t_1 < \ldots < t_m = T$, let $(\hat{S}_0, \hat\theta_0) = (S_0, \theta_0)$ and define recursively

$$
\begin{aligned}
\hat{S}_{t_{i+1}} &= \hat{S}_{t_i} + r\hat{S}_{t_i}\Delta t_{i+1} + \hat{S}_{t_i}\sqrt{\hat\theta_{t_i}\Delta t_{i+1}} \cdot Z_{i+1}, \\
\hat\theta_{t_{i+1}} &= \max\{0, \hat\theta_{t_i} + a(b - \hat\theta_{t_i})\Delta t_{i+1} + \sigma\sqrt{\hat\theta_{t_i}\Delta t_{i+1}} \cdot R_{i+1}\},
\end{aligned}
$$

for $i = 0, \ldots, m - 1$. Here $\Delta t_{i+1} = t_{i+1} - t_i$ and (Z_{i+1}, R_{i+1})'s are iid jointly normal random vectors with mean 0 and covariance matrix Σ. The plain Monte Carlo estimate for the call option price is just the sample average of iid copies of $e^{-rT}(\hat{S}_T - K)^+$. Write $R_{i+1} = \rho Z_{i+1} + \sqrt{1 - \rho^2}Y_{i+1}$, where $\{Z_1, \ldots, Z_m, Y_1, \ldots, Y_m\}$ are iid standard normal random variables. The discounted option payoff is a function of $X = (Z_1, \ldots, Z_m, Y_1, \ldots, Y_m)$, say

$$h(X) = e^{-rT}(\hat{S}_T - K)^+. \tag{9.9}$$

As for importance sampling, let the alternative sampling distribution be $N(\mu, I_{2m})$ where $\mu = (\nu, \upsilon)$ with $\nu \in \mathbb{R}^m$ and $\upsilon \in \mathbb{R}^m$. Note that we change the sampling distribution for both stock price S and volatility θ. One can use the general iterative cross-entropy method to determine a good tilting parameter (actually for moderate strike price K the basic cross-entropy scheme should suffice, analogous to Example 9.8). To find a reasonable initial tilting parameter $\hat\mu^0$, we assume it takes the form

$$\hat\mu^0 = (\hat\nu^0, \hat\upsilon^0), \quad \hat\nu^0 = x(\sqrt{\Delta t_1}, \ldots, \sqrt{\Delta t_m}), \quad \hat\upsilon^0 = -\frac{\rho}{\sqrt{1 - \rho^2}}\hat\nu^0 \tag{9.10}$$

for some $x \in \mathbb{R}$. To determine x, observe that the corresponding time discretized process $(\bar{S}, \bar\theta)$ becomes

$$
\begin{aligned}
\bar{S}_{t_{i+1}} &= \bar{S}_{t_i} + r\bar{S}_{t_i}\Delta t_{i+1} + \bar{S}_{t_i}\sqrt{\bar\theta_{t_i}\Delta t_{i+1}} \cdot \bar{Z}_{i+1}, \\
\bar\theta_{t_{i+1}} &= \max\{0, \bar\theta_{t_i} + a(b - \bar\theta_{t_i})\Delta t_{i+1} + \sigma\sqrt{\bar\theta_{t_i}\Delta t_{i+1}} \cdot \bar{R}_{i+1}\},
\end{aligned}
$$

where $\bar{X} = (\bar{Z}_1, \ldots, \bar{Z}_m, \bar{Y}_1, \ldots, \bar{Y}_m)$ is a sample from $N(\hat{\mu}^0, I_{2m})$, and

$$\bar{R}_{i+1} = \rho \bar{Z}_{i+1} + \sqrt{1 - \rho^2} \bar{Y}_{i+1}.$$

Note that $\{\bar{R}_1, \ldots, \bar{R}_m\}$ are iid *standard normal* random variables. Therefore, it is easy to see that $(\bar{S}, \bar{\theta})$ approximates (abusing notation)

$$
\begin{aligned}
dS_t &= (r + x\sqrt{\theta_t})S_t\,dt + \sqrt{\theta_t}S_t\,dW_t \\
d\theta_t &= a(b - \theta_t)\,dt + \sigma\sqrt{\theta_t}\,dB_t.
\end{aligned}
$$

We would like to pick an x so that $E[S_T]$ is approximately K. There is no explicit formula for $E[S_T]$. However, we can approximate θ_t by its long-run average b in the dynamics of S and consider instead

$$dS_t \approx (r + x\sqrt{b})S_t\,dt + \sqrt{b}S_t\,dW_t.$$

It follows that

$$E[S_T] \approx S_0 e^{(r + x\sqrt{b})T}.$$

Letting

$$x = \frac{1}{\sqrt{b}T}\log\left(\frac{K}{S_0}\right) - \frac{r}{\sqrt{b}}, \tag{9.11}$$

we arrive at $E[S_T] \approx K$. Below is the pseudocode. As before, $\hat{S}_i$ and $\hat{\theta}_i$ stand for $\hat{S}_{t_i}$ and $\hat{\theta}_{t_i}$, respectively.

Pseudocode for Heston model using the cross-entropy method:

> set x by formula (9.11)
> > initialize $\hat{v}^0 = x(\sqrt{\Delta t_1}, \ldots, \sqrt{\Delta t_m})$ and $\hat{\upsilon}^0 = -\rho\hat{v}^0/\sqrt{1 - \rho^2}$
> > set $\hat{\mu}^0 = (\hat{v}^0, \hat{\upsilon}^0)$ and set the iteration counter $j = 0$
>
> (*) generate N iid samples $\bar{X}_1, \ldots, \bar{X}_N$ from $N(\hat{\mu}^j, I_{2m})$
> > set $\hat{\mu}^{j+1} = \sum_{k=1}^N h(\bar{X}_k)e^{-\langle \hat{\mu}^j, \bar{X}_k \rangle}\bar{X}_k / \sum_{k=1}^N h(\bar{X}_k)e^{-\langle \hat{\mu}^j, \bar{X}_k \rangle}$
> > set the iteration counter $j = j + 1$
> > if $j = $ IT_NUM set $\hat{\mu} = \hat{\mu}^j$ and continue, otherwise go to step (*)
> > write $\hat{\mu} = (\hat{v}, \hat{\upsilon})$
> > for $k = 1, 2, \ldots, n$
> > > set $\hat{S}_0 = S_0$ and $\hat{\theta}_0 = \theta_0$
> > > for $i = 0, 1, \ldots, m - 1$
> > > > generate Z_{i+1} from $N(\hat{v}_{i+1}, 1)$ and Y_{i+1} from $N(\hat{\upsilon}_{i+1}, 1)$

set $R_{i+1} = \rho Z_{i+1} + \sqrt{1 - \rho^2} Y_{i+1}$

set $\hat{S}_{i+1} = \hat{S}_i + r\hat{S}_i \Delta t_{i+1} + \hat{S}_i \sqrt{\hat{\theta}_i \Delta t_{i+1}} \cdot Z_{i+1}$

set $\hat{\theta}_{i+1} = \max\{0, \hat{\theta}_i + a(b - \hat{\theta}_i)\Delta t_{i+1} + \sigma\sqrt{\hat{\theta}_i \Delta t_{i+1}} \cdot R_{i+1}\}$

compute the discounted payoff multiplied by the likelihood ratio

$$H_k = e^{-rT} (\hat{S}_m - K)^+ \cdot \exp\left\{ - \sum_{i=1}^{m}(\hat{v}_i Z_i + \hat{v}_i Y_i) + \frac{1}{2}\sum_{i=1}^{m}(\hat{v}_i^2 + \hat{v}_i^2) \right\}$$

compute the estimate $\hat{v} = \dfrac{1}{n}\sum_{i=1}^{n} H_i$

compute the standard error S.E. $= \sqrt{\dfrac{1}{n(n-1)}\left(\sum_{i=1}^{n} H_i^2 - n\hat{v}^2\right)}.$

The numerical results are reported in Table 9.8. The parameters are defined by

$$S_0 = 50, \ r = 0.03, \ b = 0.04, \ a = 1, \ \sigma = 0.1, \ \rho = 0.5,$$

$$\theta_0 = 0.05, \ m = 30, \ n = 10000.$$

The cross-entropy scheme uses $N = 2000$ pilot samples and IT_NUM $= 5$ iterations.

Table 9.8: Iterative cross-entropy method for a Heston model

	$K = 50$		$K = 70$		$K = 90$		$K = 110$	
	Plain	IS	Plain	IS	Plain	IS	Plain	IS
Estimate	5.0687	5.0495	0.5175	0.5521	0.0495	0.0514	0.0029	0.0054
S.E.	0.0819	0.0277	0.0281	0.0051	0.0097	0.0006	0.0014	0.0001

The iterative cross-entropy method significantly reduces the variance, especially when the strike price is higher.

Discussion on the initial tilting parameter $\hat{\mu}^0$: If one plays around with the given iterative cross-entropy scheme, it will soon become clear that the scheme will behave erratically for certain ranges of parameters. Sometimes this issue can be resolved by increasing the pilot sample size N or the number of iterations IT_NUM, but not always. The cause of this problem is that the initial tilting parameter given by (9.10) and (9.11) might be too far away from the optimal one. If we examine (9.10) and (9.11) more carefully, they represent an alternative sampling distribution where the distribution of the original volatility process remains the same and only the distribution of the stock price is changed. However, there are two ways for the stock price to hit the strike price K with nontrivial probability under an alternative sampling distribution: (1) the stock price is tilted higher toward K; (2) the volatility is tilted higher so that the stock price is more likely to reach K. If the optimal sampling distribution has significant contribution from the latter, the initial tilting parameter defined by (9.10) and (9.11) may not be a good choice.

A general solution is to adopt the initialization technique outlined in Section 7.2.3. The scheme is rather straightforward. Write the payoff function in (9.9) as

$$h(X) = e^{-rT}(\hat{S}_T - K)^+ = H(X; K)1_{\{F(X) \geq K\}}$$

where $F(X) = \hat{S}_T$ and $H(X; \alpha) = e^{-rT}(F(X) - \alpha)^+$. Below is the pseudocode for generating an initial tilting parameter.

Pseudocode for the initialization of the cross-entropy scheme:

    choose ρ between 5% and 10% and set $N_0 = [N(1 - \rho)]$

    set $\bar{\nu}^0 = (0, \ldots, 0)$ and $\bar{\upsilon}^0 = (0, \ldots, 0)$

    set $\bar{\mu}^0 = (\bar{\nu}^0, \bar{\upsilon}^0)$ and the iteration counter $j = 0$

(*) generate N iid samples $Y_1, \ldots, Y_N$ from $N(\bar{\mu}^j, I_{2m})$

    set $V_k = F(Y_k)$ and the order statistics $V_{(1)} \leq \cdots \leq V_{(N)}$

    set $\bar{\alpha}_{j+1} = V_{(N_0)}$

    update $\bar{\mu}^{j+1}$ according to (7.18) with $\hat{\theta}^j$ replaced by $\bar{\mu}^j$

    set the iteration counter $j = j + 1$

    if $\bar{\alpha}_j \geq K$ set $\hat{\mu}^0 = \bar{\mu}^j$ and stop, otherwise go to step (*).

Note that the final tilting parameter from this scheme will serve as the *initial tilting parameter* for the general iterative cross-entropy scheme.

Table 9.9: Cross-entropy method with general initialization technique

	$K = 130$	$b = 0.01,\ K = 100$	$\theta_0 = 0.01,\ K = 130$
Estimate	6.3417×10^{-4}	0.0064	2.5654×10^{-6}
S.E.	9.2801×10^{-6}	0.0001	4.6579×10^{-8}
NUM_IT	3	2	3

Additional numerical results are reported in Table 9.9. We have chosen those cases where the previous cross-entropy scheme [i.e., with the initial tilting parameter given by (9.10) and (9.11)] has difficulty with. The plain Euler scheme estimates are not reported as they are meaningless. The parameters of the model, except those indicated in the table, remain the same. With the general initialization technique, the iterative cross-entropy scheme is quite robust and constantly yields very accurate estimates. We should mention that the number of iterations for the general iterative cross-entropy scheme is still set at IT_NUM = 5. The entry "NUM_IT" in Table 9.9 records the number of iterations performed in the initialization program. ∎

Exercises

Pen-and-Paper Problems

9.1 Suppose that X satisfies the stochastic differential equation

$$dX_t = (2re^t \sqrt{X_t} + \sigma^2) \, dt + 2\sigma \sqrt{X_t} \, dW_t,$$

where r and σ are both positive constants.

  (a) Find a function F such that $Y_t = F(X_t)$ is a diffusion process with constant diffusion coefficient σ. That is, Y satisfies a stochastic differential equation of the form

$$dY_t = b(t, Y_t) \, dt + \sigma \, dW_t.$$

  (b) For your choice of F, compute the drift function b.

  (c) Write down a discretization scheme for X with no discretization error.

9.2 Consider a diffusion process X that satisfies the stochastic differential equation

$$dX_t = b(t, X_t) \, dt + \sigma(t, X_t) \, dW_t,$$

where $\sigma(t, x)$ is a strictly positive function. Find a function $F(t, x)$ such that $Y_t = F(t, X_t)$ satisfies

$$dY_t = \mu(t, Y_t) \, dt + dW_t$$

for some function μ. Determine F and μ. This is a generalized version of the Lamperti transform.

9.3 Find the Lamperti transform of the following diffusion processes. All the constants involved are assumed to be strictly positive.

  (a) $dX_t = a(b - X_t) \, dt + \sigma \sqrt{X_t} \, dW_t$
  (b) $dX_t = (r_0 + r_1 X_t) \, dt + \sigma X_t \, dW_t$

9.4 Suppose that r is a strictly positive constant. Find the Lamperti transform of the diffusion process X that satisfies

$$dX_t = r \, dt + \sigma X_t \, dW_t, \quad X_0 > 0.$$

Is the Lamperti transform valid when r is negative?

9.5 Write down the Euler scheme and the Milstein scheme for the following diffusion processes.

  (a) $dX_t = r \, dt + \sigma X_t \, dW_t.$
  (b) $dX_t = rX_t \, dt + \sqrt{\sigma_0^2 + \sigma_1^2 X_t^2} \, dW_t.$

9.6 Let a, b, and β be positive constants. Consider a mean-reverting Ornstein–Uhlenbeck process X such that

$$dX_t = a(b - X_t)\, dt + \beta dW_t.$$

Use the results in Example 8.6 to devise a discretization scheme for X with no discretization error.

9.7 Let $r(t)$ and $\theta(t)$ be two deterministic functions. Devise a discretization scheme for X with no discretization error, where

$$dX_t = r(t)\, X_t\, dt + \theta(t) X_t\, dW_t.$$

MATLAB® Problems

For all the numerical schemes, divide the time interval $[0, T]$ into m subintervals of equal length. That is,

$$t_i = \frac{i}{m}T, \quad i = 0, 1, \ldots, m.$$

The sample size is always denoted by n.

9.A Estimate the error $E|\hat{X}_T - X_T|$ where $\hat{X}$ is the time discretized approximating process for the diffusion

$$dX_t = a(b - X_t)\, dt + \sigma\, dW_t$$

via the Euler scheme. You will need a discretization scheme for X without discretization error (Exercise 9.6), and it will be necessary to simulate the random vector

$$\left(W_{t_{i+1}} - W_{t_i}, \int_{t_i}^{t_{i+1}} \sigma e^{at}\, dW_t \right),$$

for which you might want to use the results from Exercise 8.9. Report your results for

$$a = 2, \quad b = 1, \quad \sigma = 0.2, \quad T = 1, \quad X_0 = 1,$$

and $m = 5, 10, 20, 40, 60, 80, 100$, respectively. Let the sample size be $n = 1,000,000$. Describe the relationship between the error and the step size $h = T/m$.

9.B Assume that under the risk-neutral probability measure, the short rate r_t is a mean-reverting process:

$$dr_t = a(b - r_t)\, dt + \sigma r_t dW_t.$$

Write a function to compare the Euler scheme and the Milstein scheme in the estimation of the price of a zero-coupon bond with maturity T. Report your results for

$$r_0 = 0.12, \ a = 1, \ b = 0.10, \ \sigma = 0.2, \ T = 1, \ m = 50, \ n = 10000.$$

Is there really a difference in the performance of these two schemes?

9.C Suppose that under the risk-neutral probability measure, the stock price satisfies the stochastic differential equation

$$\frac{dS_t}{S_t} = r_t \, dt + \sigma \, dW_t$$

and the short rate $\{r_t\}$ is a Cox–Ingersoll–Ross process:

$$dr_t = a(b - r_t) \, dt + \theta \sqrt{r_t} \, dB_t,$$

where (W, B) is a two-dimensional Brownian motion with covariance matrix

$$\Sigma = \begin{bmatrix} 1 & \rho \\ \rho & 1 \end{bmatrix},$$

and a, b, σ, θ are all positive constants. Estimate the price of a call option with strike price K and maturity T by the method of conditioning. Compare with the plain Monte Carlo estimate. Report your results for

$$S_0 = 50, \ \sigma = 0.2, \ r_0 = 0.04, \ a = 1, \ b = 0.06, \ \theta = 0.1, \ \rho = 0.8, \ T = 1,$$

$$m = 50, \ n = 10000.$$

and $K = 45, 50, 55$, respectively.

9.D Assume that under the risk-neutral probability measure, the stock price S satisfies the stochastic differential equation

$$dS_t = rS_t \, dt + \sqrt{\sigma_0^2 + \sigma_1^2 S_t^2} \, dW_t,$$

where r is the risk-free interest rate and (σ_0, σ_1) are two positive constants. Estimate the prices of the following options with the control variate method and explain how you construct the control variate. Compare with the plain Monte Carlo estimate.

   (a) A call option with maturity T and strike price K.

   (b) An average price call option with maturity T and strike price K, whose payoff is

$$\left(\frac{1}{m} \sum_{i=1}^{m} S_{t_i} - K \right)^+.$$

Report your results for

$$S_0 = 50, \ r = 0.03, \ K = 50, \ T = 1, \ m = 30, \ \sigma_0 = 2, \ \sigma_1 = 0.2,$$

with sample size $n = 10000$.

9.E Repeat Exercise 9.D for the Heston model where the risk-free interest rate is a constant r and under the risk-neutral probability measure

$$
\begin{aligned}
dS_t &= rS_t \, dt + \sqrt{\theta_t} S_t \, dW_t \\
d\theta_t &= a(b - \theta_t) \, dt + \sigma\sqrt{\theta_t} \, dB_t,
\end{aligned}
$$

for some positive constants a, b, and σ such that $2ab \geq \sigma^2$. Here (W, B) is a two-dimensional Brownian motion with covariance matrix

$$\Sigma = \begin{bmatrix} 1 & \rho \\ \rho & 1 \end{bmatrix}.$$

Report your results for

$$S_0 = 50, \ r = 0.03, \ a = 1, \ b = 0.2, \ \sigma = 0.6, \ \theta_0 = 0.3, \ T = 1, \ K = 50,$$

$$\rho = 0.6, \ m = 50, n = 10000.$$

9.F Suppose that under the risk-neutral probability measure the stock price is a CEV process:

$$dS_t = rS_t \, dt + \sigma\sqrt{S_t} \, dt.$$

(a) Write a function to estimate the price of a put option with maturity T and strike price K by the basic cross-entropy method. Report your results for

$$S_0 = 50, \ r = 0.05, \ \sigma = 0.2, \ T = 1, \ m = 50, \ n = 10000,$$

and $K = 55, 52, 50$, respectively. The pilot sample size is $N = 2000$.

(b) Write a function to estimate the put option price by the general iterative cross-entropy method when the strike price K is lower. Indicate your choice of the initial tilting parameter. Report your results for $K = 48, 45$, respectively. Use IT_NUM $= 5$ iterations. All other parameters remain the same.

9.G Suppose that under the risk-neutral probability measure the stock price is a CEV process:

$$dS_t = rS_t \, dt + \sigma S_t^\gamma \, dt$$

for some $0.5 \leq \gamma < 1$. Write a function to estimate the price of an average price call option with maturity T and payoff

$$\left(\frac{1}{m} \sum_{i=1}^{m} S_{t_i} - K \right)^+,$$

by the general iterative cross-entropy method. Use IT_NUM $= 5$ iterations, each iteration using $N = 2000$ pilot samples. Report your results for

$$S_0 = 50, \ r = 0.05, \ \sigma = 0.2, \ T = 1, \ m = 50, \ n = 10000,$$

and (a) $\gamma = 0.5$, $K = 50, 52, 55$; (b) $\gamma = 0.7$, $K = 50, 55, 60$; (c) $\gamma = 0.9$, $K = 50, 60, 70$, respectively. Indicate your choice of the initial tilting parameter.

Chapter 10

Sensitivity Analysis

Sensitivity analysis is important for the purpose of hedging a portfolio or managing risk [16, 24]. Consider a simple scenario where an investment is made on a stock option with maturity T. Denote the value of the option at time $t \in [0, T]$ by

$$V(S, r, \sigma, t),$$

where S is the underlying stock price at time t, r is the risk-free interest rate, and σ is the volatility. Consider a strategy of hedging one share of the option with x shares of the stock. The value of this portfolio at time t is

$$V(S, r, \sigma, t) + xS.$$

Letting the derivative of this value with respect to the stock price S be zero, we obtain

$$x = -\frac{\partial V}{\partial S}(S, r, \sigma, t).$$

With this choice of x, one would expect the portfolio value to be insensitive to the movement of the stock price. The partial derivative

$$\frac{\partial V}{\partial S}$$

is said to be the *delta* of the option, which indicates how sensitive the value of the option is with respect to the movement of the underlying asset price. For example, if the delta of an option is 0.5, then the price of the option will roughly increase or decrease by $\$0.5$ if the stock price moves up or down by $\$1$, respectively. Similarly, if the delta of an option is -0.6, then the price of the option will roughly decrease or increase by $\$0.6$ if the stock price moves up or down by $\$1$, respectively.

Partial derivatives such as delta are referred to as *Greeks*. They measure the sensitivity of the value of a financial instrument with respect to the underlying parameters such as asset prices and volatilities. The name "Greeks" is used because these sensitivities are often denoted by Greek letters.

10.1 Commonly Used Greeks

Let $V = V(S, r, \sigma, t)$ denote the value of an option at time t, where S stands for the underlying stock price at time t, r is the risk-free interest rate, and σ is the volatility. Some commonly used Greeks are listed in Table 10.1.

Table 10.1: Commonly used Greeks

Greeks	Notation	Definition
Delta	Δ	$\partial V/\partial S$
Gamma	Γ	$\partial^2 V/\partial S^2$
Rho	ρ	$\partial V/\partial r$
Theta	Θ	$\partial V/\partial t$
Vega	ν	$\partial V/\partial \sigma$

Many times it is only one parameter, say S, that is of interest. In these cases, the option value is simply denoted by $V(S)$ if there is no confusion about other parameters.

Example 10.1. Suppose that under the risk-neutral probability measure, the underlying stock price is a geometric Brownian motion

$$S_t = S_0 \exp\left\{ \left(r - \frac{1}{2}\sigma^2\right) t + \sigma W_t \right\}.$$

Consider a call option with maturity T and strike price K. Calculate its delta at time $t = 0$.

SOLUTION: By the Black–Scholes formula, the price of the option is

$$V(S_0) = S_0 \Phi(\alpha) - Ke^{-rT}\Phi(\alpha - \sigma\sqrt{T}),$$

where

$$\alpha = \frac{1}{\sigma\sqrt{T}} \log \frac{S_0}{K} + \left(\frac{\sigma}{2} + \frac{r}{\sigma}\right)\sqrt{T}.$$

It follows from the definition of delta and straightforward calculation (we omit the details) that

$$\Delta = \frac{\partial V}{\partial S_0} = \Phi(\alpha). \tag{10.1}$$

In particular, the delta of a call option is always between 0 and 1.   ∎

10.2 Monte Carlo Simulation of Greeks

Since explicit formulas for Greeks are not available in general, Monte Carlo simulation is often used to produce estimates. Generically, one can write the Greeks in the form of derivatives. To be more concrete, consider an option with the discounted payoff $X(\theta)$, where θ is the parameter of interest. For example, if one wishes to estimate delta, then $\theta = S$; if one wishes to estimate vega, then $\theta = \sigma$. The value of the option is

$$V(\theta) = E[X(\theta)],$$

where the expected value is taken under the risk-neutral probability measure. Then estimating the Greek amounts to estimating the derivative

$$V'(\theta) = \frac{\partial}{\partial \theta} E[X(\theta)].$$

10.2.1 Methods of Finite Difference

A popular finite difference method for estimating derivatives is based on the *central difference* approximation

$$V'(\theta) = \lim_{h\downarrow 0} \frac{V(\theta + h) - V(\theta - h)}{2h}.$$

Given a small number h, let $\hat{V}(\theta + h)$ and $\hat{V}(\theta - h)$ be estimates for $V(\theta + h)$ and $V(\theta - h)$, respectively. Then an estimate for $V'(\theta)$ is

$$\frac{\hat{V}(\theta + h) - \hat{V}(\theta - h)}{2h}. \tag{10.2}$$

To reduce the variance, common random numbers are usually used to produce estimates of $\hat{V}(\theta + h)$ and $\hat{V}(\theta - h)$.

Even if $\hat{V}(\theta \pm h)$ are both unbiased, the estimate (10.2) for $V'(\theta)$ can be biased and the magnitude of the bias follows directly from the Taylor's expansion:

$$\text{bias} = \frac{V(\theta + h) - V(\theta - h)}{2h} - V'(\theta) \approx \frac{1}{6} V'''(\theta) h^2.$$

To reduce the bias, one would like to decrease h. However, the variance of the estimate

$$\frac{1}{4h^2} \text{Var}[\hat{V}(\theta + h) - \hat{V}(\theta - h)]$$

may remain bounded away from zero or even explode as $h \to 0$. Therefore, reducing h alone will not make the estimate more accurate unless one increases the sample size accordingly.

Example 10.2. Estimate the delta at $t = 0$ of a call option with maturity T and strike price K, assuming that the underlying stock price is a geometric Brownian motion with drift r (the risk-free interest rate) and volatility σ under the risk-neutral probability measure.

SOLUTION: In this case the parameter of interest is $\theta = S_0$. Given an arbitrary h, the estimate for $V(S_0 \pm h)$ is the sample average of iid copies of

$$e^{-rT} \left[(S_0 \pm h) \exp\left\{ \left(r - \frac{1}{2}\sigma^2 \right) T + \sigma\sqrt{T}Z \right\} - K \right]^+.$$

The scheme is fairly straightforward. Below is the pseudocode. Note that the same samples $\{Z_i\}$ are used for estimating $V(S_0 + h)$ and $V(S_0 - h)$.

Pseudocode of the central difference method:

> for $i = 1, 2, \ldots, n$
> > generate Z_i from $N(0,1)$
> > set $X_i = (S_0 + h) \cdot \exp\{(r - \sigma^2/2)T + \sigma\sqrt{T}Z_i\}$
> > set $Y_i = (S_0 - h) \cdot \exp\{(r - \sigma^2/2)T + \sigma\sqrt{T}Z_i\}$
> > set $H_i = \frac{1}{2h}[e^{-rT}(X_i - K)^+ - e^{-rT}(Y_i - K)^+]$
>
> compute the estimate $\hat{v} = \frac{1}{n}(H_1 + H_2 + \cdots + H_n)$
>
> compute the standard error S.E. $= \sqrt{\frac{1}{n(n-1)} \left(\sum_{i=1}^{n} H_i^2 - n\hat{v}^2 \right)}.$

The numerical results are reported in Table 10.2. The theoretical values are calculated from formula (10.1). We let

$$S_0 = 50, \quad r = 0.05, \quad \sigma = 0.2, \quad T = 1, \quad n = 10,000,000,$$

and estimate the delta for $K = 50$ and 55, respectively.

Table 10.2: Finite difference method: delta for call options

	$K = 50$ ($\Delta = 0.6368$)			$K = 55$ ($\Delta = 0.4496$)		
	$h = 1$	$h = 0.1$	$h = 0.01$	$h = 1$	$h = 0.1$	$h = 0.01$
Estimate	0.6364	0.6368	0.6366	0.4495	0.4497	0.4497
S.E.	0.0002	0.0002	0.0002	0.0002	0.0002	0.0002

In this example, the variance of the estimates remains bounded away from zero. Reducing h alone will not produce more accurate estimates if the sample size is not increased accordingly.   ∎

10.2.2   Method of Pathwise Differentiation

The basic idea of the pathwise differentiation method is to exchange the order of expectation and differentiation to obtain

$$V'(\theta) = \frac{\partial}{\partial \theta} E\left[X(\theta)\right] = E\left[\frac{\partial}{\partial \theta} X(\theta)\right].$$

The plain Monte Carlo estimate of $V'(\theta)$ is just the sample average of iid copies of

$$\frac{\partial}{\partial \theta} X(\theta),$$

which is an unbiased estimate. All the variance reduction techniques we have discussed so far can be applied to improve the efficiency of the estimate. Note that $X(\theta)$ is a random variable depending on θ. Therefore, its derivative with respect to θ is a pathwise differentiation and a random variable itself. In general, the evaluation or approximation of this derivative is rather straightforward. Many times explicit formulas are available.

Finally, it should be pointed out that the method of pathwise differentiation does not work universally. Its validity hinges on the interchangeability of the order of differentiation and expectation. The rule of thumb is that it is justifiable when the payoff function is everywhere continuous and almost everywhere continuously differentiable with respect to the parameter. One should be especially careful when the payoff is discontinuous. For example, consider the delta at $t = 0$ for a binary option with maturity T and discounted payoff

$$X(S_0) = e^{-rT}1_{\{S_T \geq K\}}.$$

Even though

$$\frac{\partial X}{\partial S_0} = 0$$

almost everywhere, the delta of this option is clearly not zero.

Example 10.3. Use the method of pathwise differentiation to estimate the delta at $t = 0$ for a call option with maturity T and strike price K, assuming that the underlying stock price is a geometric Brownian motion with drift r (the risk-free interest rate) and volatility σ under the risk-neutral probability measure.

SOLUTION: Note that $\theta = S_0$ for the calculation of delta. The stock price is

$$S_T = S_0 \exp\left\{\left(r - \frac{1}{2}\sigma^2\right)T + \sigma W_T\right\}$$

and the discounted option payoff is

$$X(S_0) = e^{-rT}(S_T - K)^+ = e^{-rT}(S_T - K)1_{\{S_T \geq K\}}.$$

It follows that

$$\frac{\partial X}{\partial S_0} = 1_{\{S_T \geq K\}}e^{-rT}\frac{\partial S_T}{\partial S_0} = 1_{\{S_T \geq K\}}\frac{e^{-rT}S_T}{S_0}$$

and

$$\Delta = E\left[1_{\{S_T \geq K\}}\frac{e^{-rT}S_T}{S_0}\right].$$

Monte Carlo simulation of Δ is straightforward. The numerical results are reported in Table 10.3 with parameters

$$S_0 = 50, \quad r = 0.05, \quad \sigma = 0.2, \quad T = 1, \quad n = 10000.$$

The theoretical values of delta are obtained from formula (10.1).   ∎

Table 10.3: Pathwise differentiation: delta for call options

Strike price K	40	45	50	55	60
Theoretical value	0.9286	0.8097	0.6368	0.4496	0.2872
Estimate	0.9244	0.8064	0.6373	0.4469	0.2867
S.E.	0.0036	0.0049	0.0058	0.0059	0.0054

Example 10.4. The setup is the same as Example 10.3. Consider a discretely monitored average price call option with maturity T and payoff

$$(\bar{S} - K)^+, \quad \bar{S} = \frac{1}{m} \sum_{i=1}^{m} S_{t_i},$$

where $0 = t_0 < t_1 < \cdots < t_m = T$ are prefixed dates. Argue that the delta of this option at $t = 0$ is

$$E\left[1_{\{\bar{S} \geq K\}} \frac{e^{-rT} \bar{S}}{S_0}\right].$$

See Exercise 10.5 for a generalization of this result.

PROOF: The discounted payoff of this option is $X(S_0) = e^{-rT}(\bar{S} - K)^+$. By the method of pathwise differentiation, its delta is $\Delta = E[\partial X / \partial S_0]$. Similar to Example 10.3,

$$\begin{aligned}
\frac{\partial X}{\partial S_0} &= 1_{\{\bar{S} \geq K\}} \cdot e^{-rT} \frac{\partial \bar{S}}{\partial S_0} \\
&= 1_{\{\bar{S} \geq K\}} \cdot e^{-rT} \frac{1}{m} \sum_{i=1}^{m} \frac{\partial S_{t_i}}{\partial S_0} \\
&= 1_{\{\bar{S} \geq K\}} \cdot e^{-rT} \frac{1}{m} \sum_{i=1}^{m} \frac{S_{t_i}}{S_0} \\
&= 1_{\{\bar{S} \geq K\}} \cdot e^{-rT} \frac{\bar{S}}{S_0}.
\end{aligned}$$

We complete the proof. ■

Example 10.5. Control variate method by means of artificial dynamics. Consider a stochastic volatility model where under the risk-neutral probability measure

$$\begin{aligned}
dS_t &= rS_t \, dt + \theta_t S_t \, dW_t, \\
d\theta_t &= a(\Theta - \theta_t) \, dt + \beta \, dB_t.
\end{aligned}$$

Here r is the risk-free interest rate, (a, Θ, β) are all positive constants, and (W, B) is a two-dimensional Brownian motion with covariance matrix

$$\Sigma = \begin{bmatrix} 1 & \rho \\ \rho & 1 \end{bmatrix}.$$

Estimate the delta of a call option with maturity T and strike price K.

SOLUTION: This example is very similar to Example 9.7. It follows from Itô formula that

$$S_t = S_0 \exp\{Y_t\}, \tag{10.3}$$

where Y_t satisfies

$$dY_t = \left(r - \frac{1}{2}\theta_t^2 \right) dt + \theta_t \, dW_t, \quad Y_0 = 0.$$

Since the process Y does *not* depend on S_0, the delta of the call option is

$$\Delta = E\left[1_{\{S_T \geq K\}} e^{-rT} \frac{\partial S_T}{\partial S_0} \right] = E\left[1_{\{S_T \geq K\}} e^{-rT} \frac{S_T}{S_0} \right].$$

Plugging in (10.3), we have

$$\Delta = E\left[1_{\{Y_T \geq y\}} \exp\{Y_T - rT\} \right], \quad y = \log\left(\frac{K}{S_0} \right).$$

It is straightforward to apply the Euler scheme on the two-dimensional process (Y, θ) to estimate Δ. More precisely, let $0 = t_0 < t_1 < \cdots < t_m = T$ be a time discretization. Define $\hat{Y}_0 = Y_0 = 0$ and $\hat{\theta}_0 = \theta_0$. For $i = 0, 1, \ldots, m-1$, recursively define

$$\hat{Y}_{t_{i+1}} = \hat{Y}_{t_i} + \left(r - \frac{1}{2}\hat{\theta}_{t_i}^2 \right)(t_{i+1} - t_i) + \hat{\theta}_{t_i} \sqrt{t_{i+1} - t_i} Z_{i+1},$$

$$\hat{\theta}_{t_{i+1}} = \hat{\theta}_{t_i} + a(\Theta - \hat{\theta}_{t_i})(t_{i+1} - t_i) + \beta \sqrt{t_{i+1} - t_i} R_{i+1},$$

where $\{(Z_1, R_1), \ldots, (Z_m, R_m)\}$ are iid jointly normal random vectors with distribution $N(0, \Sigma)$. The plain Monte Carlo estimate for Δ is just the sample average of iid copies of

$$X = 1_{\{\hat{Y}_T \geq y\}} \exp\{\hat{Y}_T - rT\}.$$

Similar to Example 9.7, one can introduce an *artificial* stochastic process $\bar{Y}$ defined recursively by $\bar{Y}_0 = Y_0 = 0$ and

$$\bar{Y}_{t_{i+1}} = \bar{Y}_{t_i} + \left(r - \frac{1}{2}\sigma^2 \right)(t_{i+1} - t_i) + \sigma \sqrt{t_{i+1} - t_i} Z_{i+1}$$

for $i = 0, 1, \ldots, m - 1$. This process $\bar{Y}$ uses the *same* sequence $\{Z_1, \ldots, Z_m\}$ and some *constant* volatility σ. The control variate is defined to be

$$V = 1_{\{\bar{Y}_T \geq y\}} \exp\{\bar{Y}_T - rT\}.$$

Clearly, Y_T would equal $\bar{Y}_T$ if the volatility θ_t were the constant σ. This implies that $E[V]$ is the delta of a call option with the same maturity T and strike price K when the underlying stock price is a geometric Brownian motion with drift r and volatility σ. In particular, by formula (10.1),

$$E[V] = \Phi(\alpha), \quad \alpha = \frac{1}{\sigma\sqrt{T}} \log \frac{S_0}{K} + \left(\frac{\sigma}{2} + \frac{r}{\sigma}\right) \sqrt{T}. \tag{10.4}$$

The control variate estimate for the delta is the sample average of iid copies of

$$X - b[V - \Phi(\alpha)]$$

for some constant b. As in Example 9.7, we choose $\sigma = \Theta$ because Θ is the long run average of $\{\theta_t\}$ and use $b = \hat{b}^*$, the sample estimate for the optimal coefficient b^*; see formula (6.2).

Below is the pseudocode. As before, we use $\hat{Y}_i$, $\bar{Y}_i$, and $\hat{\theta}_i$ to denote $\hat{Y}_{t_i}$, $\bar{Y}_{t_i}$, and $\hat{\theta}_{t_i}$, respectively.

Pseudocode for control variate method:

> set $\sigma = \Theta$, $y = \log(K/S_0)$, and α as in (10.4)
> for $k = 1, 2, \ldots, n$
> > set $\hat{Y}_0 = 0$, $\bar{Y}_0 = 0$, and $\hat{\theta}_0 = \theta_0$
> > for $i = 0, 1, \ldots, m - 1$
> > > generate iid sample Z_{i+1} and U_{i+1} from $N(0,1)$
> > > set $R_{i+1} = \rho Z_{i+1} + \sqrt{1 - \rho^2} U_{i+1}$
> > > set $\hat{Y}_{i+1} = \hat{Y}_i + (r - \hat{\theta}_i^2/2)(t_{i+1} - t_i) + \hat{\theta}_i \sqrt{t_{i+1} - t_i} Z_{i+1}$
> > > set $\bar{Y}_{i+1} = \bar{Y}_i + (r - \sigma^2/2)(t_{i+1} - t_i) + \sigma\sqrt{t_{i+1} - t_i} Z_{i+1}$
> > > set $\hat{\theta}_{i+1} = \hat{\theta}_i + a(\Theta - \hat{\theta}_i)(t_{i+1} - t_i) + \beta\sqrt{t_{i+1} - t_i} R_{i+1}$
> > set $X_k = 1_{\{\hat{Y}_T \geq y\}} \exp\{\hat{Y}_T - rT\}$;
> > set $Q_k = 1_{\{\bar{Y}_T \geq y\}} \exp\{\bar{Y}_T - rT\} - \Phi(\alpha)$
> compute $\hat{b}^*$ from formula (6.2) [with Y replaced by Q]
> for $k = 1, 2, \ldots, n$
> > set $H_k = X_k - \hat{b}^* Q_k$
> compute the estimate $\hat{v} = \frac{1}{n} \sum_{i=1}^{n} H_i$

$$\text{compute the standard error S.E.} = \sqrt{\frac{1}{n(n-1)}\left(\sum_{i=1}^{n} H_i^2 - n\hat{v}^2\right)}.$$

We compare the plain Monte Carlo scheme with the control variate method. The numerical results are reported in table 10.4 for

$$S_0 = 50,\ r = 0.03,\ T = 1,\ a = 2,\ \beta = 0.1,\ \theta_0 = 0.25,\ \Theta = 0.2,$$

$$\rho = 0.5,\ m = 50,\ t_i = iT/m,\ n = 10000.$$

Table 10.4: Estimating Δ: control variate method

	$K = 45$		$K = 50$		$K = 55$	
	Plain	Control	Plain	Control	Plain	Control
Estimate	0.7218	0.7221	0.5978	0.6030	0.4816	0.4839
S.E.	0.0062	0.0032	0.0066	0.0026	0.0067	0.0023

The control variate method does produce more accurate estimates. More extensive numerical experiments should show that the choice of $\sigma = \Theta$ is nearly optimal. ∎

Example 10.6. Delta by Euler Scheme. Assume that the risk-free interest rate r is a constant and the underlying stock price S satisfies the stochastic differential equation

$$dS_t = rS_t\,dt + \sigma(S_t)\,dW_t$$

under the risk-neutral probability measure. Design an estimate for the delta at $t = 0$ of an option with maturity T and payoff $h(S_T)$.

Solution: The price of the option is $V(S_0) = E[e^{-rT}h(S_T)]$. Therefore, by the chain rule,

$$\Delta = E\left[e^{-rT}\frac{\partial}{\partial S_0}h(S_T)\right] = E\left[e^{-rT}h'(S_T)\frac{\partial S_T}{\partial S_0}\right].$$

However, unlike all the examples we have seen so far, it is not possible to explicitly calculate

$$\frac{\partial S_T}{\partial S_0} \tag{10.5}$$

in general. To approximate it, consider the Euler scheme for S. Given a time discretization $0 = t_0 < t_1 < \cdots < t_m = T$. Define recursively

$$\hat{S}_0 = S_0, \tag{10.6}$$

$$\hat{S}_{t_{i+1}} = \hat{S}_{t_i} + r\hat{S}_{t_i}(t_{i+1} - t_i) + \sigma(\hat{S}_{t_i})\sqrt{t_{i+1} - t_i}\, Z_{i+1} \tag{10.7}$$

for $i = 0, 1, \ldots, m-1$, where $\{Z_1, \ldots, Z_m\}$ are iid standard normal random variables. Define

$$\hat{\Delta}_{t_i} = \frac{\partial \hat{S}_{t_i}}{\partial S_0}, \quad i = 0, 1, \ldots, m.$$

Taking derivatives over S_0 on both sides of (10.6) and (10.7), it follows from the chain rule that

$$\hat{\Delta}_0 = 1, \tag{10.8}$$

$$\hat{\Delta}_{t_{i+1}} = \hat{\Delta}_{t_i} + r\hat{\Delta}_{t_i}(t_{i+1} - t_i) + \sigma'(\hat{S}_{t_i})\hat{\Delta}_{t_i}\sqrt{t_{i+1} - t_i}\, Z_{i+1}. \tag{10.9}$$

Equations (10.6)–(10.9) offer a recursive algorithm to compute the pathwise derivative

$$\hat{\Delta}_{t_m} = \hat{\Delta}_T = \frac{\partial \hat{S}_T}{\partial S_0},$$

which can be used as an approximation of (10.5). The plain Monte Carlo estimate of the delta is simply the sample average of iid copies

$$e^{-rT} h'(\hat{S}_T)\hat{\Delta}_T.$$

Note that the above calculation can be extended to path-dependent options and more general price dynamics. We leave this as an exercise to the interested reader. ∎

10.2.3   Method of Score Function

Contrary to the method of pathwise differentiation, the method of score function, sometimes also referred to as the *likelihood ratio method*, does not require any smoothness condition on the payoff function.

Recall that $X(\theta)$ stands for the discounted payoff of the option and the option price $V(\theta)$ is given by

$$V(\theta) = E[X(\theta)].$$

Often, one can write $X(\theta) = H(Y)$ for some function H that has *no* dependence on θ, and some random variable or random vector Y that has a θ-dependent density $f_\theta(y)$. It follows that

$$V(\theta) = E[H(Y)] = \int H(y) f_\theta(y) \, dy.$$

Taking derivate with respect to θ and exchanging the order of integration and differentiation, we arrive at

$$V'(\theta) = \int H(y) \frac{\partial f_\theta}{\partial \theta}(y) \, dy = \int H(y) \frac{\partial \log f_\theta}{\partial \theta}(y) \cdot f_\theta(y) \, dy,$$

or equivalently,

$$V'(\theta) = E\left[H(Y) \frac{\partial \log f_\theta}{\partial \theta}(Y)\right] = E\left[X(\theta) \frac{\partial \log f_\theta}{\partial \theta}(Y)\right].$$

Therefore, an unbiased estimate for $V'(\theta)$ is just the sample mean of iid copies of

$$X(\theta) \frac{\partial \log f_\theta}{\partial \theta}(Y).$$

Note that the particular form of H is not important. The essential requirement is that the discounted payoff X can be written as a θ-independent function of some random variable or random vector Y. There are many possible ways to choose Y. But usually Y is chosen to be the building block of the sampling scheme for $X(\theta)$ or one with the simplest form of density.

Observe the difference between the pathwise differentiation method and the score function method. In the former, the differentiation is taken with respect to the payoff function, while in the latter, the differentiation is taken with respect to the density function of some underlying random variable or random vector. For the score function method to work, it requires that the density of Y be smooth with respect to θ, which is much milder than the regularity condition on $X(\theta)$ for the pathwise differentiation method. Another advantage of this method is that once the *score function*

$$\frac{\partial \log f_\theta}{\partial \theta}$$

is obtained, it can be applied to any payoff function X. However, the practical use of this method is often limited by the explicit knowledge of the density function f_θ.

We should illustrate the score function method through examples. In all these examples, we assume that the risk-free interest rate r is a constant and the underlying stock price is a geometric Brownian motion with drift r and volatility σ under the risk-neutral probability measure. All the Greeks under consideration are assumed to be at $t = 0$. For Greeks at a general time point $t < T$, it suffices to replace T by $T - t$ and S_0 by S_t.

Example 10.7. Write down an estimate for the delta of an option with maturity T and payoff $h(S_T)$.

SOLUTION: The parameter of interest is $\theta = S_0$. The stock price at maturity T is

$$S_T = S_0 \exp\left\{ \left(r - \frac{1}{2}\sigma^2 \right) T + \sigma\sqrt{T}Z \right\},$$

where Z is a standard normal random variable. Define

$$Y = Z + \frac{1}{\sigma\sqrt{T}} \log S_0.$$

Then the option price is

$$V(S_0) = E[H(Y)]$$

where H is a function independent of $\theta = S_0$:

$$H(Y) = e^{-rT} h(S_T) = e^{-rT} h\left(\exp\left\{ \left(r - \frac{1}{2}\sigma^2 \right) T + \sigma\sqrt{T}Y \right\} \right)$$

The distribution of Y is normal and depends on S_0. Its density, denoted by f_{S_0}, is

$$f_{S_0}(y) = \frac{1}{\sqrt{2\pi}} \exp\left\{ -\frac{1}{2}\left(y - \frac{1}{\sigma\sqrt{T}} \log S_0 \right)^2 \right\},$$

which satisfies

$$\frac{\partial \log f_{S_0}}{\partial S_0}(y) = \frac{y}{S_0\sigma\sqrt{T}} - \frac{1}{S_0\sigma^2 T} \log S_0.$$

It follows that

$$\frac{\partial \log f_{S_0}}{\partial S_0}(Y) = \frac{Y}{S_0\sigma\sqrt{T}} - \frac{1}{S_0\sigma^2 T} \log S_0 = \frac{Z}{S_0\sigma\sqrt{T}},$$

which in turn implies that

$$\Delta = E\left[H(Y)\frac{\partial \log f_{S_0}}{\partial S_0}(Y) \right] = E\left[e^{-rT} h(S_T)\frac{Z}{S_0\sigma\sqrt{T}} \right].$$

An unbiased estimate for Δ is just the sample average of iid copies of the random variable inside expectation. In this formulation, h does not need to be continuous, for example, h can be the payoff of a binary option.   ■

Example 10.8. Write down an estimate for the vega of an option with maturity T and payoff $h(S_T)$.

SOLUTION: The parameter of interest is $\theta = \sigma$. The stock price at maturity T is

$$S_T = S_0 \exp\left\{\left(r - \frac{1}{2}\sigma^2\right)T + \sigma\sqrt{T}Z\right\},$$

where Z is a standard normal random variable. Define

$$Y = \log\left(\frac{S_T}{S_0}\right) - rT = \sigma\sqrt{T}Z - \frac{1}{2}\sigma^2 T. \tag{10.10}$$

Then the price of the option is

$$V(\sigma) = E[H(Y)],$$

where

$$H(Y) = e^{-rT}h(S_T) = e^{-rT}h(S_0 \cdot \exp\{Y + rT\}).$$

The distribution of Y depends on $\theta = \sigma$. Indeed, Y is normally distributed with distribution

$$N\left(-\frac{1}{2}\sigma^2 T, \sigma^2 T\right),$$

and its density, denoted by f_σ, satisfies

$$\frac{\partial \log f_\sigma}{\partial \sigma}(y) = \frac{y^2}{\sigma^3 T} - \frac{\sigma T}{4} - \frac{1}{\sigma}.$$

Therefore,

$$\begin{aligned}
\frac{\partial \log f_\sigma}{\partial \sigma}(Y) &= \frac{Y^2}{\sigma^3 T} - \frac{\sigma T}{4} - \frac{1}{\sigma} \\
&= \frac{1}{\sigma^3 T}\left(\sigma\sqrt{T}Z - \frac{1}{2}\sigma^2 T\right)^2 - \frac{\sigma T}{4} - \frac{1}{\sigma} \\
&= \frac{Z^2 - 1}{\sigma} - \sqrt{T}Z. \tag{10.11}
\end{aligned}$$

It follows that

$$v = E\left[H(Y)\frac{\partial \log f_\sigma}{\partial \sigma}(Y)\right] = E\left[e^{-rT}h(S_T)\left(\frac{Z^2 - 1}{\sigma} - \sqrt{T}Z\right)\right].$$

An unbiased estimate for vega is just the sample average of iid copies of the random variable inside expectation. We want to repeat that h does not need to be continuous. ∎

Example 10.9. Write down an estimate for the vega of an average price call option with maturity T and payoff

$$(\bar{S} - K)^+, \quad \bar{S} = \frac{1}{m} \sum_{i=1}^{m} S_{t_i},$$

where $0 = t_0 < t_1 < \cdots < t_m = T$ are given dates.

SOLUTION: In this case $\theta = \sigma$. Denote $\Delta t_{i+1} = t_{i+1} - t_i$. Then we can recursively write

$$S_{t_{i+1}} = S_{t_i} \exp \left\{ \left(r - \frac{1}{2}\sigma^2 \right) \Delta t_{i+1} + \sigma \sqrt{\Delta t_{i+1}} Z_{i+1} \right\},$$

for $i = 0, 1, \ldots, m-1$. Similar to (10.10), define for each i

$$Y_{i+1} = \log \left(\frac{S_{t_{i+1}}}{S_{t_i}} \right) - r\Delta t_{i+1} = \sigma \sqrt{\Delta t_{i+1}} Z_{i+1} - \frac{1}{2}\sigma^2 \Delta t_{i+1}.$$

Since $S_{t_{i+1}} = S_{t_i} \exp\{r\Delta t_{i+1} + Y_{i+1}\}$ for each i, it follows easily that

$$e^{-rT}(\bar{S} - K)^+ = H(Y_1, \ldots, Y_m)$$

for some function H that does not depend on σ. Because Y_i's are independent, the joint density of $(Y_1, \ldots, Y_m)$, denoted by $f_\sigma(y_1, \ldots, y_m)$, satisfies

$$f_\sigma(y_1, \ldots, y_m) = \prod_{i=1}^{m} f_\sigma^{(i)}(y_i),$$

where $f_\sigma^{(i)}$ is the marginal density for Y_i. Therefore, thanks to (10.11),

$$\frac{\partial \log f_\sigma}{\partial \sigma}(Y_1, \ldots, Y_m) = \sum_{i=1}^{m} \frac{\partial \log f_\sigma^{(i)}}{\partial \sigma}(Y_i) = \sum_{i=1}^{m} \left(\frac{Z_i^2 - 1}{\sigma} - \sqrt{\Delta t_i} Z_i \right).$$

It follows that

$$\begin{aligned}
\nu &= E\left[H(Y_1, \ldots, Y_m) \frac{\partial \log f_\sigma}{\partial \sigma}(Y_1, \ldots, Y_m) \right] \\
&= E\left[e^{-rT}(\bar{S} - K)^+ \sum_{i=1}^{m} \left(\frac{Z_i^2 - 1}{\sigma} - \sqrt{\Delta t_i} Z_i \right) \right].
\end{aligned}$$

In the preceding derivation, the specific nature of the average price call option is not important at all. As long as the option payoff is of the form $h(S_{t_1}, \ldots, S_{t_m})$, the formula for vega is still valid if one replaces the payoff of the average price call by h. ∎

Exercises

Unless otherwise specified, we assume that the risk-free interest rate r is a constant and the underlying stock price is a geometric Brownian motion with drift r and volatility σ under the risk-neutral probability measure. All Greeks under consideration are assumed to be at $t = 0$.

Pen-and-Paper Problems

10.1 Explicitly evaluate the delta and vega of a put option with maturity T and strike price K.

10.2 Explicitly evaluate the delta of a binary option with maturity T and payoff

$$1_{\{S_T \geq K\}}.$$

10.3 Use the method of pathwise differentiation to show that the vega of a call option with maturity T and strike price K is

$$v = E\left[1_{\{S_T \geq K\}} e^{-rT} S_T \left(\frac{1}{\sigma} \log \frac{S_T}{S_0} - \left(\frac{r}{\sigma} + \frac{\sigma}{2}\right) T\right)\right].$$

10.4 Use the method of pathwise differentiation to find an unbiased estimate for the delta of a discretely monitored lookback call option with maturity T and payoff

$$S_T - \min_{1 \leq i \leq m} S_{t_i},$$

where $0 = t_0 < t_1 < \cdots < t_m = T$ are given dates.

10.5 Assume that the stock price S satisfies the stochastic differential equation

$$dS_t = r_t S_t \, dt + \theta_t S_t \, dW_t,$$

under the risk-neutral probability measure. Here r_t and θ_t are the short rate and volatility at time t, respectively. They can be deterministic or stochastic, but are assumed to have no dependence on S_0. Consider an option with maturity T and payoff $h(S_{t_0}, S_{t_1}, \ldots, S_{t_m})$, where $0 = t_0 < t_1 < \cdots < t_m = T$ are give dates. Show that the delta of this option is

$$\Delta = E\left[\sum_{i=0}^{m} \exp\left\{-\int_0^T r_t \, dt\right\} \frac{S_{t_i}}{S_0} \frac{\partial}{\partial S_{t_i}} h(S_{t_0}, S_{t_1}, \ldots, S_{t_m})\right].$$

10.6 Consider the following generalization of Example 10.6. Let X be a diffusion process that satisfies the stochastic differential equation

$$dX_t = b(X_t) \, dt + \sigma(X_t) \, dW_t.$$

Define $V(X_0) = E[h(X_{t_0}, X_{t_1}, \ldots, X_{t_m})]$, where $0 = t_0 < t_1 < \cdots < t_m = T$ are given time points. Design an Euler scheme to approximate

$$\frac{\partial V}{\partial X_0}.$$

You can assume that h is nice so that the order of differentiation and expectation can be exchanged.

10.7 Mimic Example 10.9 to write down an unbiased estimate for the delta of a path-dependent option with maturity T and payoff $h(S_{t_1}, \ldots, S_{t_m})$, where $0 = t_0 < t_1 < \cdots < t_m = T$ are given dates. Use the score function method.

10.8 Suppose that we are interested in an unbiased estimate for the delta and vega of a down-and-in barrier option with maturity T and payoff

$$(S_T - K)^+ 1_{\{\min(S_{t_1}, \ldots, S_{t_m}) \leq b\}},$$

where $0 = t_0 < t_1 < \cdots < t_m = T$ are given dates. Use the score function method to design your estimates. Will the pathwise differentiation method work? Why?

MATLAB® Problems

10.A Use the central difference method to estimate the following Greeks. The sample size is always $n = 100000$, and the model parameters are

$$S_0 = 50, \quad r = 0.03, \quad \sigma = 0.2, \quad T = 1.$$

(a) Delta of a put option with maturity T and strike price K. Report your results for $K = 50$ and $h = 1, 0.1, 0.01$, respectively.

(b) Vega of a call option with maturity T and strike price K. Report you results for $K = 50$ and $h = 0.1, 0.01, 0.001$, respectively.

(c) Delta of a up-and-out barrier option with maturity T and payoff

$$(S_T - K)^+ 1_{\{\max(S_{t_1}, \ldots, S_{t_m}) \leq b\}}.$$

Report your results for $K = 50$, $b = 65$, $m = 30$, $t_i = iT/m$, and $h = 10^{-k}$ with $k = 0, 1, \ldots, 6$, respectively. What is your conclusion?

10.B Consider an average price call option with maturity T and strike price K, whose payoff is

$$(\bar{S} - K)^+, \quad \bar{S} = \frac{1}{m} \sum_{i=1}^{m} S_{t_i}.$$

Write a function to estimate the delta by

(a) the method of pathwise differentiation.

(b) the method of pathwise differentiation, combined with the control variate method to reduce variance. Use the delta of the corresponding average price call option with geometric mean as the control variate. Use the sample estimate of b^*. *Hint:* For an average price call option with geometric mean, show that its delta is

$$\Delta = E\left[1_{\{\bar{S}_G \geq K\}}\frac{e^{-rT}\bar{S}_G}{S_0}\right], \quad \bar{S}_G = \left(\prod_{i=1}^{m} S_{t_i}\right)^{\frac{1}{m}},$$

and derive the theoretical delta value from the explicit pricing formula for the average price call option with geometric mean.

Report your estimates and standard errors for

$$S_0 = 50, \ r = 0.03, \ K = 50, \ \sigma = 0.2, \ T = 1, \ m = 30, \ t_i = iT/m,$$

with sample size $n = 10000$.

10.C Suppose that under the risk-neutral probability measure, the stock price satisfies the stochastic differential equation:

$$\begin{aligned} dS_t &= r_t S_t \, dt + \sigma S_t \, dW_t \\ dr_t &= a(b - r_t) \, dt + \theta r_t \, dB_t. \end{aligned}$$

Here a, b, σ, θ are all positive constants, and (W, B) is a two-dimensional Brownian motion with covariance matrix

$$\Sigma = \begin{bmatrix} 1 & \rho \\ \rho & 1 \end{bmatrix}.$$

Let $Y_t = \log S_t$. Consider a call option with maturity T and strike price K.

(a) Apply Exercise 10.5 to express Δ in terms of an expected value.

(b) Estimate Δ by the Euler scheme on (Y_t, r_t).

(c) Mimic Example 9.6 to estimate Δ by the method of conditioning.

Report your estimates and standard errors for

$$S_0 = 50, \ r_0 = 0.03, \ a = 1, \ b = 0.02, \ \sigma = 0.2, \ \theta = 0.1, \ T = 1,$$

$$K = 50, \ m = 50, \ t_i = iT/m, \ n = 10000,$$

and $\rho = 0.1, 0.3, 0.5, 0.7, 0.9$, respectively. Explain why the variance reduction becomes less significant when ρ gets larger.

10.D Assume that the risk-free interest rate r is a constant and under the risk-neutral probability measure the underlying stock price satisfies the stochastic differential equation

$$dS_t = rS_t \, dt + \sigma \sqrt{S_t} \, dW_t.$$

Design an Euler scheme to estimate the following Greeks.

(a) The delta of a call option with maturity T and strike price K.

(b) The delta of an average price call option with maturity T and strike price K, whose payoff is

$$(\bar{S} - K)^+, \quad \bar{S} = \frac{1}{m} \sum_{i=1}^{m} S_{t_i}.$$

You will need a slight modification on the Euler scheme to ensure that the time discretized process $\hat{S}$ stays nonnegative. Modify the Euler scheme on Δ accordingly. Report your results for

$$S_0 = 50, \quad r = 0.03, \quad \sigma = 0.3, \quad K = 50, \quad T = 1,$$

$$m = 30, \quad t_i = iT/m, \quad n = 10000.$$

10.E Use the score function method to estimate the delta and vega of a binary option with maturity T and payoff $1_{\{S_T \geq K\}}$. Report your results for

$$S_0 = 50, \quad r = 0.03, \quad K = 50, \quad \sigma = 0.2, \quad T = 1, \quad n = 10000.$$

Compare with the theoretical values.

10.F Use the score function method to estimate the delta and vega of an up-and-out barrier option with maturity T and payoff

$$(S_T - K)^+ 1_{\{\max(S_{t_1},...,S_{t_m}) \leq b\}}.$$

Report your results for

$$S_0 = 50, \quad r = 0.05, \quad \sigma = 0.2, \quad T = 1, \quad m = 30, \quad t_i = iT/m,$$

$$K = 50, \quad b = 70, \quad n = 10000.$$

Appendix A

Multivariate Normal Distributions

In this appendix, we review some of the important results concerning jointly normal random vectors and multivariate normal distributions.

The convention in this appendix is that the vectors are always in *column* form. Given a random vector $X = (X_1, \ldots, X_n)'$, its *expected value* is defined to be the vector

$$E[X] = (E[X_1], \ldots, E[X_n])'.$$

Let $Y = (Y_1, \ldots, Y_m)'$ be another random vector. The *covariance matrix* of X and Y is defined to be the $n \times m$ matrix

$$\text{Cov}(X, Y) = [\text{Cov}(X_i, Y_j)].$$

That is, the entry of the covariance matrix in the i-th row and j-th column is $\text{Cov}(X_i, Y_j)$. In particular, the $n \times n$ symmetric matrix $\text{Var}[X] = \text{Cov}(X, X)$ is said to be the covariance matrix of X.

A random vector $X = (X_1, X_2, \ldots, X_n)'$ is said to be *jointly normal* and have a *multivariate normal distribution* $N(\mu, \Sigma)$, if the joint density function of X takes the form

$$f(x) = (2\pi)^{-\frac{n}{2}} |\Sigma|^{-\frac{1}{2}} e^{-\frac{1}{2}(x-\mu)' \Sigma^{-1} (x-\mu)}, \quad x \in \mathbb{R}^n.$$

Here $\mu = (\mu_1, \ldots, \mu_n)'$ is an n-dimensional vector and $\Sigma = [\Sigma_{ij}]$ is an $n \times n$ symmetric positive definite matrix whose determinant is denoted by $|\Sigma|$. In the case where $\mu = 0$ and $\Sigma = I_n$ ($n \times n$ identity matrix), we say the distribution is the *n-dimensional standard normal*.

Marginal density functions

Since a collection of continuous random variables are independent if and only if the joint density function is the product of individual marginal density functions, we have the following result immediately.

Lemma A.1. $X = (X_1, \ldots, X_n)'$ *is an n-dimensional standard normal random vector if and only if* $X_1, \ldots, X_n$ *are independent standard normal random variables. More generally, if* $X_1, \ldots, X_n$ *are independent normal variables with* $E[X_i] = \mu_i$ *and* $Var[X_i] = \sigma_i^2$, *then* $X = (X_1, \ldots, X_n)'$ *is a jointly normal random vector with distribution* $N(\mu, \Sigma)$ *where*

$$\mu = \begin{bmatrix} \mu_1 \\ \mu_2 \\ \vdots \\ \mu_n \end{bmatrix}, \quad \Sigma = \begin{bmatrix} \sigma_1^2 & 0 & \cdots & 0 \\ 0 & \sigma_2^2 & \cdots & 0 \\ \vdots & \vdots & \ddots & \vdots \\ 0 & 0 & \cdots & \sigma_n^2 \end{bmatrix}.$$

The converse is also true.

We will list some of the properties of multivariate normal distributions without proof. For a comprehensive treatment, see [32]. Assume from now on that $X = (X_1, \ldots, X_n)'$ is jointly normal with distribution $N(\mu, \Sigma)$.

a. $E[X] = \mu$ and $Var[X] = \Sigma$.

b. X_i has distribution $N(\mu_i, \Sigma_{ii})$ for each $i = 1, \ldots, n$.

c. X_i and X_j are independent if and only if $\Sigma_{ij} = Cov(X_i, X_j) = 0$.

d. For any $m \times n$ matrix C and $m \times 1$ vector b, the random vector

$$Y = CX + b$$

is jointly normal with distribution $N(C\mu + b, C\Sigma C')$.

e. Given any $k < n$, let $Y_1 = (X_1, \ldots, X_k)'$ and $Y_2 = (X_{k+1}, \ldots, X_n)'$. Consider the partition of X, μ, and Σ given by

$$X = \begin{bmatrix} Y_1 \\ Y_2 \end{bmatrix}, \quad \mu = \begin{bmatrix} \theta_1 \\ \theta_2 \end{bmatrix}, \quad \Sigma = \begin{bmatrix} \Lambda_{11} & \Lambda_{12} \\ \Lambda_{21} & \Lambda_{22} \end{bmatrix},$$

where

$$\theta_i = E[Y_i], \quad \Lambda_{ij} = Cov(Y_i, Y_j).$$

Then the conditional distribution of Y_1 given $Y_2 = y$ is $N(\bar{\mu}, \bar{\Sigma})$ with

$$\bar{\mu} = \theta_1 + \Lambda_{12}\Lambda_{22}^{-1}(y - \theta_2), \quad \bar{\Sigma} = \Lambda_{11} - \Lambda_{12}\Lambda_{22}^{-1}\Lambda_{21}.$$

Appendix B

American Option Pricing

Throughout the book, it has been assumed that an option can only be exercised at maturity. Such options are called *European options*. The mechanism of an *American option* is no different, except that the holder of the option can exercise the option at any time before or at maturity. Therefore, an American option is at least as valuable as its European counterpart.

In a nutshell, evaluating an American option amounts to solving a control problem—at any given time, there are two actions (i.e., controls) that are available to the option holder, namely, to exercise the option immediately or to wait. If the value of exercising is larger than that of waiting, the option holder should exercise, and vice versa. While it is trivial that the value of exercising equals the option payoff, the difficult part of this decision process is to determine the value of waiting.

A systematic approach for solving such control problems is *dynamic programming*. It is a recursive algorithm running *backwards* in time. The main goal of the appendix is to discuss this approach and use it to price American options via binomial tree approximations. Due to the complexity of evaluating American options in general, we refer the reader to [11] for an in-depth discussion on various different techniques.

B.1   The Value of an American Option

Consider an American option with maturity T and payoff X_t if the option is exercised at time $t \leq T$. For example, for an American put option with strike price K

$$X_t = (K - S_t)^+.$$

The value of this option can be expressed in terms of the solution to an optimization problem:

$$v = \max_{0 \leq \tau \leq T} E[e^{-r\tau} X_\tau], \qquad (B.1)$$

where the maximum is taken over all meaningful random times τ. Not surprisingly, the expected value is taken under the risk-neutral probability measure.

We will first discuss what are "meaningful" random times. Roughly speaking, a meaningful random time is an exercise time one can realize *without* looking into the future. For instance, a strategy of exercising the option whenever the stock price reaches down to a certain level b corresponds to the random time

$$\tau = \min\{t \in [0, T] : S_t \leq b\};$$

a strategy of exercising the option at a prefixed time t_0 no matter how the stock price behaves corresponds to

$$\tau = t_0.$$

These two examples of random times or strategies are meaningful and practical. On the contrary, a strategy of exercising the option when the stock price reaches the absolute minimum on the time interval $[0, T]$ corresponds to

$$\tau = \min\{t \in [0, T] : S_t = \min_{0 \leq u \leq T} S_u\},$$

which is neither practical nor meaningful. It will require the option holder to know the *entire* path of the stock price—whether the absolute minimum is reached or not cannot be determined unless at maturity. This is clearly unrealistic since the option holder, making a decision at time t, can only use the information he has gathered up until time t. From these discussions, it is not surprising that meaningful random times are also said to be *nonanticipating*.

The pricing formula (B.1) can be interpreted as follows. If the option is to be exercised at a meaningful random time τ, the value of the option at $t = 0$ is

$$E[e^{-r\tau} X_\tau],$$

where the expected value is taken under the risk-neutral probability measure. Since the option can be exercised at any meaningful random time τ, it will be natural for the option holder to maximize the above expected value. Therefore, the value of the option at $t = 0$ should be

$$\max_{0 \leq \tau \leq T} E[e^{-r\tau} X_\tau],$$

and the optimal strategy for the option holder is to exercise the option at a random time τ^* that solves the above optimization problem.

Remark B.1. From now on, whenever we say random times, we implicitly assume that they are meaningful or nonanticipating. Such random times are also called *stopping times* in the literature, and this type of specialized control problems (B.1) are often said to be "optimal stopping problems."

B.2   Dynamic Programming and Binomial Trees

Evaluating an American option amounts to solving a stochastic optimization problem. The principle of dynamic programming essentially breaks it down into a sequence of much simpler, static optimization problems that can be solved backwards in time. We should illustrate this approach through the binomial tree pricing models.

The setup is exactly the same as that of Section 3.2.3. Consider a binomial tree with N periods. At each time period, the stock price either moves up by a factor u or moves down by a factor d, and the discounting factor is $1/R$. Under the risk-neutral probability measure, the stock price moves up with probability p^* and moves down with probability q^*, where

$$p^* = \frac{R - d}{u - d}, \quad q^* = 1 - p^* = \frac{u - R}{u - d}.$$

We will index the time steps by $n = 0, 1, \ldots, N$ and let S_n denote the stock price at time n. Consider an American option with maturity N and payoff, if exercised at time n,

$$H(S_n).$$

To solve this American option pricing problem, let $V_n(x)$ be the value of the option at time n if $S_n = x$. The price of the American option at time 0 is, by definition,

$$v = V_0(S_0).$$

By introducing these unknown functions $\{V_n(x) : n = 0, 1 \ldots, N\}$, it seems that we have made the matter more complicated. However, this allows us to write down a recursive equation (i.e., *dynamic programming equation*) in the following fashion: for $n = 0, 1, \ldots, N-1$,

$$V_n(x) = \max\{H(x), E[V_{n+1}(S_{n+1})|S_n = x]/R\}, \qquad (B.2)$$
$$V_N(x) = H(x). \qquad (B.3)$$

Equation (B.3) is said to be the *terminal condition*. It is self-explanatory in the sense that the value of the option equals its payoff at maturity. The recursive equation (B.2) amounts to that the value of an American option at time n (given $S_n = x$) equals the maximum of the value of exercising, which is option payoff $H(x)$, and the value of waiting, which is $E[V_{n+1}(S_{n+1})|S_n = x]/R$. In the setting of binomial trees, the value of waiting is easy to evaluate and

$$E[V_{n+1}(S_{n+1})|S_n = x] = p^* V_{n+1}(ux) + q^* V_{n+1}(dx).$$

These considerations lead to a simple recursive algorithm for computing the value of the American option at any time. It is very similar to European option pricing with binomial trees: (1) compute the value of the option at time N by the terminal condition; (2) recursively compute the value of the option at each node by the dynamic programming equation, backwards in time. This recursive algorithm also yields the optimal exercise strategy: at any node, it is optimal to exercise if the value of exercising is greater than the value of waiting, and vice versa.

Example B.1. Consider an American put option with maturity $T = 1$ year and strike price $K = 50$. The underlying stock price is assumed to be a geometric Brownian motion with initial price $S_0 = 50$ and volatility $\sigma = 0.2$. The risk-free interest is $r = 12\%$ annually. Approximate the price of this American put option with a binomial tree of three periods.

SOLUTION: There are $N = 3$ periods. Each period represents $\Delta t = T/N = 1/3$ year in real time. Therefore, the parameters of the approximating binomial tree are

$$R = e^{r\Delta t} = 1.0408, \quad u = e^{\sigma\sqrt{\Delta t}} = 1.1224, \quad d = e^{-\sigma\sqrt{\Delta t}} = 0.8909,$$

and the risk-neutral probabilities are

$$p^* = \frac{R-d}{u-d} = 0.6475, \quad q^* = \frac{u-R}{u-d} = 0.3525.$$

Figure B.1 includes the tree of stock price, the tree of option value, and the tree of optimal strategy. For example, consider the node where the stock price is 39.6894. For this node, the value of exercising is

$$(50 - 39.6894)^+ = 10.3106$$

and the value of waiting is

$$(p^* \cdot 5.4526 + q^* \cdot 14.6389)/R = 8.3501.$$

Therefore, the value of the option at this node is 10.3106 and the optimal strategy is to immediately exercise. The price of the option at time 0 is 2.2359. ∎

Lemma B.1. *The value of an American call option equals the value of an European call option with the same maturity and strike price.*

PROOF. We only need to show that for an American call option it is always optimal to wait until maturity to exercise. Denote by $V_n(x)$ the value of the American call option at time n if $S_n = x$. It follows from the dynamic program equation (B.2) that, for $n = 0, 1, \ldots, N - 1$,

$$V_n(x) = \max\{(x - K)^+, \ E[V_{n+1}(S_{n+1})|S_n = x]/R\}.$$

Using the trivial observation $V_{n+1}(x) \geq (x - K)^+$ and the inequality $x^+ + y^+ \geq (x + y)^+$ for any x and y, we arrive at

$$
\begin{aligned}
E[V_{n+1}(S_{n+1})|S_n = x] \ &= \ p^* V_{n+1}(ux) + q^* V_{n+1}(dx) \\
&\geq \ p^*(ux - K)^+ + q^*(dx - K)^+ \\
&\geq \ (p^* ux + q^* dx - K)^+ \\
&= \ (Rx - K)^+ \\
&\geq \ R(x - K)^+,
\end{aligned}
$$

which in turn implies that

$$V_n(x) = E[V_{n+1}(S_{n+1})|S_n = x]/R.$$

In other words, it is always optimal to wait at time n for $n = 0, 1, \ldots, N - 1$. This completes the proof. ∎

B.3 Diffusion Models: Binomial Approximation

When the underlying asset price is a diffusion, it is possible to approximate it with binomial trees. We have already seen the use of such approximation to geometric Brownian motions. This section will discuss the extension to general diffusions.

Recall the binomial tree approximation of a geometric Brownian motion S that satisfies

$$\frac{dS_t}{S_t} = r\,dt + \sigma\,dW_t.$$

It starts with a single node S_0. In each period it moves up by a factor u with probability p^* or moves down by a factor d with probability $1 - p^*$, where

$$u = e^{\sigma\sqrt{\Delta t}}, \ d = e^{-\sigma\sqrt{\Delta t}}, \ p^* = \frac{e^{r\Delta t} - d}{u - d},$$

and Δt represents the real-time increment of each period of the binomial tree. Now suppose that we take logarithm of the stock price at each node of the binomial tree. The resulting new binomial tree starts at $\log S_0$ and, in each period, moves up by $h = \log u = \sigma\sqrt{\Delta t}$ with probability p^* or moves down by $-h = \log d$ with probability $1 - p^*$, where

$$p^* = \frac{e^{r\Delta t} - d}{u - d} = \frac{e^{r\Delta t} - e^{-h}}{e^h - e^{-h}}.$$

Obviously, this tree is an approximation to $X_t = \log S_t$, which satisfies

$$dX_t = \left(r - \frac{1}{2}\sigma^2\right)dt + \sigma\,dW_t.$$

Note that the binomial tree approximation of a geometric Brownian motion is constructed by matching the first and second moments of the increments of the process. Since such a construction is purely local, it is not difficult to generalize to a binomial tree approximation of a more general diffusion process (abusing notation) X_t that satisfies

$$dX_t = \left[b(X_t) - \frac{1}{2}\sigma^2(X_t)\right]dt + \sigma(X_t)\,dW_t.$$

More precisely, the approximating binomial tree will start at X_0. Given that its value at the current node is x, at the next time step, it will move up by $h(x)$ with probability $p^*(x)$ and move down by $-h(x)$ with probability $1 - p^*(x)$, where

$$h(x) = \sigma(x)\sqrt{\Delta t}, \quad p^*(x) = \frac{e^{b(x)\Delta t} - e^{-h(x)}}{e^{h(x)} - e^{-h(x)}}. \tag{B.4}$$

This is indeed a legitimate approximation in theory under mild regularity conditions. However, it has little use in practice because the increment $h(x)$ depends on x, and hence the nodes will not collapse as in the case of a geometric Brownian motion. Indeed, for a tree with N periods, the total number of nodes can be

$$1 + 2 + \cdots + 2^N = 2^{N+1} - 1,$$

whereas the approximating tree to a geometric Brownian motion will only have

$$1 + 2 + \cdots + (N + 1) = (N + 1)(N + 2)/2$$

nodes. A simple solution to this issue is to make $h(x)$, or equivalently, $\sigma(x)$, independent of x. This motivates the use of the Lamperti transform (see Section 9.4) to convert the original process into a diffusion with constant volatility and then apply the binomial tree approximation given by (B.4). We will illustrate this approach through an example.

Example B.2. Suppose that the risk-free interest rate r is a constant and under the risk-neutral probability measure the underlying stock price satisfies the stochastic differential equation

$$dS_t = rS_t\, dt + \sigma\sqrt{S_t}\, dW_t.$$

Use binomial tree approximation to estimate the price of an American put option with maturity T and strike price K.

SOLUTION: For this process the Lamperti transform is

$$F(x) = \int \frac{1}{\sigma\sqrt{x}}\, dx = \frac{2}{\sigma}\sqrt{x}, \quad X_t = F(S_t) = \frac{2}{\sigma}\sqrt{S_t}.$$

It follows from Itô formula that

$$dX_t = \left(\frac{rX_t}{2} - \frac{1}{2X_t}\right) dt + dW_t = \left[b(X_t) - \frac{1}{2}\right] dt + dW_t,$$

where

$$b(x) = \frac{rx}{2} - \frac{1}{2x} + \frac{1}{2}.$$

The parameters for the binomial tree approximation of X is given by (B.4), that is,

$$h = \sqrt{\Delta t}, \quad p^*(x) = \frac{e^{b(x)\Delta t} - e^{-h}}{e^h - e^{-h}}.$$

Provided that the value of X at the current node is x, it will go up to $x + h$ with probability $p^*(x)$ and drop down to $x - h$ with probability $1 - p^*(x)$.

This description of the binomial tree is not entirely accurate, however. Some slight modifications are necessary.

1. To make sure that $p^*(x)$ is a real probability, that is, $0 \le p^*(x) \le 1$, we actually let

$$p^*(x) = \min\left\{ 1, \left(\frac{e^{b(x)\Delta t} - e^{-h}}{e^h - e^{-h}} \right)^+ \right\}. \tag{B.5}$$

2. Since X is nonnegative, whenever the value of a node reaches below zero, we reset it to zero.

3. When $x = 0$, $p^*(x)$ is not well defined. However, observe that $S_t = 0$ if $X_t = 0$, and that the stock price will remain at zero once it reaches zero. Therefore, it is natural to let the approximating binomial tree stay at zero once it reaches zero. In other words, there are no more bifurcations once the value of a node reaches zero.

These modifications can be easily implemented and the resulting binomial tree is still a valid approximation to the process X.

The corresponding dynamic programming equation is straightforward. Suppose that we divide the time interval $[0, T]$ into N equal-length subintervals and let $t_n = nT/N$ for $n = 0, 1, \ldots, N$. Let $\Delta t = T/N, h = \sqrt{\Delta t}$, and define $p^*(x)$ by (B.5) for $x > 0$. Since $S_t = \sigma^2 X_t^2 / 4$, we define

$$H(x) = \left(K - \frac{1}{4}\sigma^2 x^2 \right)^+,$$

which is the option payoff given $X_t = x$.

Denote by $V_n(x)$ the value of the American option at time t_n when $X_{t_n} = x$. The dynamic programming equation is simply

$$V_N(x) = H(x),$$
$$V_n(x) = \max\left\{H(x),\, e^{-r\Delta t}[p^*(x)V_{n+1}(x+h) + q^*(x)V_{n+1}(x-h)]\right\},$$

where $q^*(x) = 1 - p^*(x)$. As we have mentioned, (i) when $x - h < 0$, we simply replace it by zero; (ii) when $x = 0$, the recursive equation becomes

$$V_n(0) = \max\left\{H(0), e^{-r\Delta t}V_{n+1}(0)\right\},$$

since the binomial tree will stay at zero.

Table B.1 presents some numerical results for the price of an American put option with maturity T and strike price K, where $S_0 = 50$, $r = 0.05$, $\sigma = 1.0$, and $T = 0.5$.

Table B.1: American option pricing with binomial trees.

	K = 45			K = 50		
	N = 50	N = 200	N = 500	N = 50	N = 200	N = 500
Price	0.2464	0.2442	0.2444	1.5251	1.5300	1.5310

Note that the binomial tree approximation is not a Monte Carlo simulation technique. Therefore, there is no standard errors associated with these prices. ∎

Remark B.2. The binomial tree approximation described here is a special case of the so-called Markov chain approximation to diffusion processes. It is a powerful numerical technique for solving general continuous time stochastic control problems. There are a myriad of ways to construct implementable approximations to diffusion processes, and the Lamperti transform is not really necessary. The interested reader may consult [21].

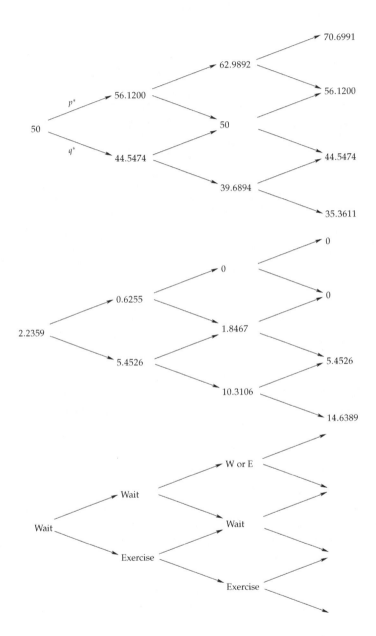

Figure B.1: American option pricing with binomial trees.

Appendix C

Option Pricing Formulas

In this appendix, we collect a number of explicit option pricing formulas. Unless otherwise specified, the underlying stock price is assumed to be a geometric Brownian motion under the risk-neutral probability measure:

$$S_t = S_0 \exp\left\{\left(r - \frac{1}{2}\sigma^2\right)t + \sigma W_t\right\},$$

where W is a standard Brownian motion, r is the risk-free interest rate, and σ is the volatility. All options are assumed to have maturity T. The option payoff is denoted by X.

i. **Binary call option:** $X = 1_{\{S_T \geq K\}}$.

$$\text{Price} = e^{-rT} \Phi\left(-\frac{\log(K/S_0) - (r - \sigma^2/2)T}{\sigma\sqrt{T}}\right).$$

ii. **Binary put option:** $X = 1_{\{S_T \leq K\}}$.

$$\text{Price} = e^{-rT} \Phi\left(\frac{\log(K/S_0) - (r - \sigma^2/2)T}{\sigma\sqrt{T}}\right).$$

iii. **Call option:** $X = (S_T - K)^+$.

$$\text{Price} = S_0 \Phi(\sigma\sqrt{T} + \theta) - Ke^{-rT}\Phi(\theta),$$

$$\theta = \frac{1}{\sigma\sqrt{T}} \log\frac{S_0}{K} + \left(\frac{r}{\sigma} - \frac{\sigma}{2}\right)\sqrt{T}.$$

iv. **Put option:** $X = (K - S_T)^+$.

$$\text{Price} = K e^{-rT} \Phi(\theta) - S_0 \Phi(\theta - \sigma \sqrt{T}),$$

$$\theta = \frac{1}{\sigma \sqrt{T}} \log \frac{K}{S_0} + \left(\frac{\sigma}{2} - \frac{r}{\sigma} \right) \sqrt{T}.$$

v. **Average price call option with geometric mean:** $X = (\bar{S}_G - K)^+$

with $\bar{S}_G = \left(\prod_{k=1}^{m} S_{t_k} \right)^{1/m}$.

$$\text{Price} = e^{-rT} \left[e^{\bar{\mu} + \frac{1}{2} \bar{\sigma}^2} \Phi(\bar{\sigma} + \theta) - K \Phi(\theta) \right],$$

$$\bar{\mu} = \log S_0 + \frac{1}{m} \left(r - \frac{\sigma^2}{2} \right) \sum_{k=1}^{m} t_k, \quad \bar{\sigma} = \frac{\sigma}{m} \sqrt{\sum_{k=1}^{m} (2m - 2k + 1) t_k},$$

$$\theta = \frac{\bar{\mu} - \log K}{\bar{\sigma}}.$$

vi. **Average price put option with geometric mean:** $X = (K - \bar{S}_G)^+$

with $\bar{S}_G = \left(\prod_{k=1}^{m} S_{t_k} \right)^{1/m}$.

$$\text{Price} = e^{-rT} \left[K \Phi(\theta) - e^{\bar{\mu} + \frac{1}{2} \bar{\sigma}^2} \Phi(\theta - \bar{\sigma}) \right],$$

$$\bar{\mu} = \log S_0 + \frac{1}{m} \left(r - \frac{\sigma^2}{2} \right) \sum_{k=1}^{m} t_k, \quad \bar{\sigma} = \frac{\sigma}{m} \sqrt{\sum_{k=1}^{m} (2m - 2k + 1) t_k},$$

$$\theta = \frac{\log K - \bar{\mu}}{\bar{\sigma}}.$$

vii. **Lookback call option with floating strike:** $X = S_T - \min_{0 \le t \le T} S_t$.

$$\text{Price} = S_0 \Phi(\theta) - \frac{S_0 \sigma^2}{2r} \Phi(-\theta) - S_0 e^{-rT} \left(1 - \frac{\sigma^2}{2r} \right) \Phi(\theta - \sigma \sqrt{T}),$$

$$\theta = \left(\frac{r}{\sigma} + \frac{\sigma}{2} \right) \sqrt{T}.$$

viii. Lookback put option with floating strike: $X = \max\limits_{0 \leq t \leq T} S_t - S_T.$

$$\text{Price} = S_0 e^{-rT}\left(1 - \frac{\sigma^2}{2r}\right)\Phi(\sigma\sqrt{T} - \theta) - S_0\Phi(-\theta) + \frac{S_0\sigma^2}{2r}\Phi(\theta),$$

$$\theta = \left(\frac{r}{\sigma} + \frac{\sigma}{2}\right)\sqrt{T}.$$

ix. Lookback call option with fixed strike: $X = \left(\max\limits_{0 \leq t \leq T} S_t - K\right)^+.$

(a) If $S_0 < K$:

$$\begin{aligned}
\text{Price} \;=\;& S_0\Phi(\theta_+) - Ke^{-rT}\Phi(\theta_+ - \sigma\sqrt{T}) \\
&+ \frac{\sigma^2}{2r}S_0\left[\Phi(\theta_+) - e^{-rT}\left(\frac{K}{S_0}\right)^{2r/\sigma^2}\Phi(\theta_-)\right],
\end{aligned}$$

$$\theta_\pm = \frac{1}{\sigma\sqrt{T}}\log\frac{S_0}{K} + \left(\frac{\sigma}{2} \pm \frac{r}{\sigma}\right)\sqrt{T}.$$

(b) If $S_0 \geq K$:

$$\text{Price} = S_0\Phi(\theta_+)\left(1 + \frac{\sigma^2}{2r}\right) + S_0 e^{-rT}\Phi(\theta_-)\left(1 - \frac{\sigma^2}{2r}\right) - Ke^{-rT},$$

$$\theta_\pm = \left(\frac{\sigma}{2} \pm \frac{r}{\sigma}\right)\sqrt{T}.$$

x. Lookback put option with fixed strike: $X = \left(K - \min\limits_{0 \leq t \leq T} S_t\right)^+.$

(a) If $S_0 < K$:

$$\text{Price} = Ke^{-rT} + S_0 e^{-rT}\Phi(\theta_-)\left(\frac{\sigma^2}{2r} - 1\right) - S_0\Phi(\theta_+)\left(1 + \frac{\sigma^2}{2r}\right),$$

$$\theta_\pm = -\left(\frac{\sigma}{2} \pm \frac{r}{\sigma}\right)\sqrt{T}.$$

(b) If $S_0 \geq K$:

$$\begin{aligned}
\text{Price} \;=\;& e^{-rT}K\Phi(\theta_+ + \sigma\sqrt{T}) - S_0\Phi(\theta_+) \\
&+ \frac{\sigma^2}{2r}S_0\left[e^{-rT}\left(\frac{K}{S_0}\right)^{2r/\sigma^2}\Phi(\theta_-) - \Phi(\theta_+)\right],
\end{aligned}$$

$$\theta_\pm = \frac{1}{\sigma\sqrt{T}}\log\frac{K}{S_0} - \left(\frac{\sigma}{2} \pm \frac{r}{\sigma}\right)\sqrt{T}.$$

xi. Down-and-out call option: $X = (S_T - K)^+ 1_{\{\min_{0 \le t \le T} S_t \ge b\}}$ with $S_0 > b$.

(a) If $b \le K$:

$$\text{Price} = S_0 \Phi(\theta) - Ke^{-rT} \Phi(\theta - \sigma\sqrt{T}) - b \left(\frac{b}{S_0}\right)^{2r/\sigma^2} \Phi(\mu)$$

$$+ e^{-rT} \frac{KS_0}{b} \left(\frac{b}{S_0}\right)^{2r/\sigma^2} \Phi(\mu - \sigma\sqrt{T}),$$

$$\theta = \frac{1}{\sigma\sqrt{T}} \log \frac{S_0}{K} + \left(\frac{\sigma}{2} + \frac{r}{\sigma}\right)\sqrt{T}, \quad \mu = \theta + \frac{2}{\sigma\sqrt{T}} \log \frac{b}{S_0}.$$

(b) If $b > K$:

$$\text{Price} = S_0 \Phi(\theta_+) - Ke^{-rT} \Phi(\theta_+ - \sigma\sqrt{T}) - b \left(\frac{b}{S_0}\right)^{2r/\sigma^2} \Phi(\theta_-)$$

$$+ e^{-rT} \frac{KS_0}{b} \left(\frac{b}{S_0}\right)^{2r/\sigma^2} \Phi(\theta_- - \sigma\sqrt{T}),$$

$$\theta_\pm = \left(\frac{r}{\sigma} + \frac{\sigma}{2}\right)\sqrt{T} \pm \frac{1}{\sigma\sqrt{T}} \log \frac{S_0}{b}.$$

xii. Down-and-in call option: $X = (S_T - K)^+ 1_{\{\min_{0 \le t \le T} S_t < b\}}$ with $S_0 > b$.

(a) If $b \le K$:

$$\text{Price} = b \left(\frac{b}{S_0}\right)^{2r/\sigma^2} \Phi(\theta) - e^{-rT} \frac{KS_0}{b} \left(\frac{b}{S_0}\right)^{2r/\sigma^2} \Phi(\theta - \sigma\sqrt{T}),$$

$$\theta = \frac{1}{\sigma\sqrt{T}} \log \frac{b^2}{KS_0} + \left(\frac{\sigma}{2} + \frac{r}{\sigma}\right)\sqrt{T}.$$

(b) If $b > K$:

$$\text{Price} = S_0 \left[\Phi(\mu) - \Phi(\theta_+)\right] - Ke^{-rT} \left[\Phi(\mu - \sigma\sqrt{T}) - \Phi(\theta_+ - \sigma\sqrt{T})\right]$$

$$+ \left(\frac{b}{S_0}\right)^{2r/\sigma^2} \left[b\Phi(\theta_-) - e^{-rT} \frac{KS_0}{b} \Phi(\theta_- - \sigma\sqrt{T})\right],$$

$$\theta_\pm = \left(\frac{r}{\sigma} + \frac{\sigma}{2}\right)\sqrt{T} \pm \frac{1}{\sigma\sqrt{T}} \log \frac{S_0}{b}, \quad \mu = \theta_+ + \frac{1}{\sigma\sqrt{T}} \log \frac{b}{K}.$$

xiii. Up-and-out call option: $X = (S_T - K)^+ 1_{\{\max_{0 \le t \le T} S_t \le b\}}$ with $S_0 < b$.

(a) If $b \le K$:

$$\text{Price} = 0.$$

(b) If $b > K$:

$$
\begin{aligned}
\text{Price} = \ & S_0 \left[\Phi(\mu) - \Phi(\theta_+) \right] - Ke^{-rT} \left[\Phi(\mu - \sigma\sqrt{T}) - \Phi(\theta_+ - \sigma\sqrt{T}) \right] \\
& + b \left(\frac{b}{S_0} \right)^{2r/\sigma^2} \left[\Phi(\nu - \sigma\sqrt{T}) - \Phi(-\theta_-) \right] \\
& - e^{-rT} \frac{KS_0}{b} \left(\frac{b}{S_0} \right)^{2r/\sigma^2} \left[\Phi(\nu) - \Phi(-\theta_- + \sigma\sqrt{T}) \right],
\end{aligned}
$$

$$
\theta_{\pm} = \left(\frac{r}{\sigma} + \frac{\sigma}{2} \right) \sqrt{T} \pm \frac{1}{\sigma\sqrt{T}} \log \frac{S_0}{b}, \quad \mu = \theta_+ + \frac{1}{\sigma\sqrt{T}} \log \frac{b}{K},
$$

$$
\nu = \frac{1}{\sigma\sqrt{T}} \log \frac{KS_0}{b^2} + \left(\frac{\sigma}{2} - \frac{r}{\sigma} \right) \sqrt{T}.
$$

xiv. Up-and-in call option: $X = (S_T - K)^+ 1_{\{\max_{0 \le t \le T} S_t > b\}}$ with $S_0 < b$.

(a) If $b \le K$:

$$\text{Price} = S_0 \Phi(\theta) - Ke^{-rT} \Phi(\theta - \sigma\sqrt{T}),$$

$$\theta = \frac{1}{\sigma\sqrt{T}} \log \frac{S_0}{K} + \left(\frac{\sigma}{2} + \frac{r}{\sigma} \right) \sqrt{T}.$$

(b) If $b > K$:

$$
\begin{aligned}
\text{Price} = \ & S_0 \Phi(\theta_+) - Ke^{-rT} \Phi(\theta_+ - \sigma\sqrt{T}) \\
& - b \left(\frac{b}{S_0} \right)^{2r/\sigma^2} \left[\Phi(\nu - \sigma\sqrt{T}) - \Phi(-\theta_-) \right] \\
& + e^{-rT} \frac{KS_0}{b} \left(\frac{b}{S_0} \right)^{2r/\sigma^2} \left[\Phi(\nu) - \Phi(-\theta_- + \sigma\sqrt{T}) \right],
\end{aligned}
$$

$$
\theta_{\pm} = \left(\frac{r}{\sigma} + \frac{\sigma}{2} \right) \sqrt{T} \pm \frac{1}{\sigma\sqrt{T}} \log \frac{S_0}{b}, \quad \nu = \frac{1}{\sigma\sqrt{T}} \log \frac{KS_0}{b^2} + \left(\frac{\sigma}{2} - \frac{r}{\sigma} \right) \sqrt{T}.
$$

xv. Down-and-out put option: $X = (K - S_T)^+ \mathbf{1}_{\{\min_{0 \le t \le T} S_t \ge b\}}$ **with** $S_0 > b$.

(a) If $b \le K$:

$$
\begin{aligned}
\text{Price} \;=\;& S_0 \left[\Phi(\mu) - \Phi(\theta_+) \right] - Ke^{-rT} \left[\Phi(\mu - \sigma\sqrt{T}) - \Phi(\theta_+ - \sigma\sqrt{T}) \right] \\
&+ b \left(\frac{b}{S_0} \right)^{2r/\sigma^2} \left[\Phi(\nu - \sigma\sqrt{T}) - \Phi(-\theta_-) \right] \\
&- e^{-rT} \frac{KS_0}{b} \left(\frac{b}{S_0} \right)^{2r/\sigma^2} \left[\Phi(\nu) - \Phi(-\theta_- + \sigma\sqrt{T}) \right],
\end{aligned}
$$

$$
\theta_\pm = \left(\frac{r}{\sigma} + \frac{\sigma}{2} \right) \sqrt{T} \pm \frac{1}{\sigma\sqrt{T}} \log \frac{S_0}{b}, \quad \mu = \theta_+ + \frac{1}{\sigma\sqrt{T}} \log \frac{b}{K},
$$

$$
\nu = \frac{1}{\sigma\sqrt{T}} \log \frac{KS_0}{b^2} + \left(\frac{\sigma}{2} - \frac{r}{\sigma} \right) \sqrt{T}.
$$

(b) If $b > K$:

$$
\text{Price} = 0.
$$

xvi. Down-and-in put option: $X = (K - S_T)^+ \mathbf{1}_{\{\min_{0 \le t \le T} S_t < b\}}$ **with** $S_0 > b$.

(a) If $b \le K$:

$$
\begin{aligned}
\text{Price} \;=\;& -S_0 \Phi(-\theta_+) + Ke^{-rT} \Phi(\sigma\sqrt{T} - \theta_+) \\
&- b \left(\frac{b}{S_0} \right)^{2r/\sigma^2} \left[\Phi(\nu - \sigma\sqrt{T}) - \Phi(-\theta_-) \right] \\
&+ e^{-rT} \frac{KS_0}{b} \left(\frac{b}{S_0} \right)^{2r/\sigma^2} \left[\Phi(\nu) - \Phi(-\theta_- + \sigma\sqrt{T}) \right],
\end{aligned}
$$

$$
\theta_\pm = \left(\frac{r}{\sigma} + \frac{\sigma}{2} \right) \sqrt{T} \pm \frac{1}{\sigma\sqrt{T}} \log \frac{S_0}{b}, \quad \nu = \frac{1}{\sigma\sqrt{T}} \log \frac{KS_0}{b^2} + \left(\frac{\sigma}{2} - \frac{r}{\sigma} \right) \sqrt{T}.
$$

(b) If $b > K$:

$$
\text{Price} = Ke^{-rT} \Phi(\theta) - S_0 \Phi(\theta - \sigma\sqrt{T}),
$$

$$
\theta = \frac{1}{\sigma\sqrt{T}} \log \frac{K}{S_0} + \left(\frac{\sigma}{2} - \frac{r}{\sigma} \right) \sqrt{T}.
$$

xvii. Up-and-out put option: $X = (K - S_T)^+ 1_{\{\max_{0 \le t \le T} S_t \le b\}}$ with $S_0 < b$.

(a) If $b \le K$:

$$
\begin{aligned}
\text{Price} \ = \ & -S_0 \Phi(-\theta_+) + K e^{-rT} \Phi(\sigma\sqrt{T} - \theta_+) + b \left(\frac{b}{S_0}\right)^{2r/\sigma^2} \Phi(-\theta_-) \\
& - e^{-rT} \frac{K S_0}{b} \left(\frac{b}{S_0}\right)^{2r/\sigma^2} \Phi(\sigma\sqrt{T} - \theta_-),
\end{aligned}
$$

$$
\theta_\pm = \left(\frac{r}{\sigma} + \frac{\sigma}{2}\right)\sqrt{T} \pm \frac{1}{\sigma\sqrt{T}} \log \frac{S_0}{b}.
$$

(b) If $b > K$:

$$
\begin{aligned}
\text{Price} \ = \ & -S_0 \Phi(-\theta) + K e^{-rT} \Phi(\sigma\sqrt{T} - \theta) + b \left(\frac{b}{S_0}\right)^{2r/\sigma^2} \Phi(-\mu) \\
& - e^{-rT} \frac{K S_0}{b} \left(\frac{b}{S_0}\right)^{2r/\sigma^2} \Phi(\sigma\sqrt{T} - \mu),
\end{aligned}
$$

$$
\theta = \frac{1}{\sigma\sqrt{T}} \log \frac{S_0}{K} + \left(\frac{\sigma}{2} + \frac{r}{\sigma}\right)\sqrt{T}, \quad \mu = \theta + \frac{2}{\sigma\sqrt{T}} \log \frac{b}{S_0}.
$$

xviii. Up-and-in put option: $X = (K - S_T)^+ 1_{\{\max_{0 \le t \le T} S_t > b\}}$ with $S_0 < b$.

(a) If $b \le K$:

$$
\begin{aligned}
\text{Price} \ = \ & S_0 \left[\Phi(\mu) - \Phi(\theta_+)\right] - K e^{-rT} \left[\Phi(\mu - \sigma\sqrt{T}) - \Phi(\theta_+ - \sigma\sqrt{T})\right] \\
& - \left(\frac{b}{S_0}\right)^{2r/\sigma^2} \left[b\Phi(-\theta_-) - e^{-rT} \frac{K S_0}{b} \Phi(\sigma\sqrt{T} - \theta_-)\right],
\end{aligned}
$$

$$
\theta_\pm = \left(\frac{r}{\sigma} + \frac{\sigma}{2}\right)\sqrt{T} \pm \frac{1}{\sigma\sqrt{T}} \log \frac{S_0}{b}, \quad \mu = \theta_+ + \frac{1}{\sigma\sqrt{T}} \log \frac{b}{K}.
$$

(a) If $b > K$:

$$
\text{Price} = -b \left(\frac{b}{S_0}\right)^{2r/\sigma^2} \Phi(-\theta) + e^{-rT} \frac{K S_0}{b} \left(\frac{b}{S_0}\right)^{2r/\sigma^2} \Phi(\sigma\sqrt{T} - \theta),
$$

$$
\theta = \frac{1}{\sigma\sqrt{T}} \log \frac{b^2}{K S_0} + \left(\frac{\sigma}{2} + \frac{r}{\sigma}\right)\sqrt{T}.
$$

xix. **Exchange options:** $X = [S_T^{(1)} - S_T^{(2)}]^+$. Assume that under the risk-neutral probability measure, the prices of the two underlying assets are

$$S_t^{(1)} = S_0^{(1)} \exp\left\{ \left(r - \frac{1}{2}\sigma_1^2 \right) t + \sigma_1 W_t^{(1)} \right\},$$

$$S_t^{(2)} = S_0^{(2)} \exp\left\{ \left(r - \frac{1}{2}\sigma_2^2 \right) t + \sigma_2 W_t^{(2)} \right\},$$

where $(W^{(1)}, W^{(2)})$ is a two-dimensional Brownian motion with co-variance matrix

$$\begin{bmatrix} 1 & \rho \\ \rho & 1 \end{bmatrix}.$$

Then

$$\text{Price} = S_0^{(1)} \Phi(\theta) - S_0^{(2)} \Phi(\theta - \sigma\sqrt{T}),$$

$$\theta = \frac{1}{\sigma\sqrt{T}} \log \frac{S_0^{(1)}}{S_0^{(2)}} + \frac{\sigma\sqrt{T}}{2}, \quad \sigma = \sqrt{\sigma_1^2 + \sigma_2^2 - 2\rho\sigma_1\sigma_2}.$$

Bibliography

[1] M. Baxter and A. Rennie. *Financial Calculus*. Cambridge University Press, Cambridge, UK, 2000.

[2] P. Billingsley. *Probability and Measure*. John Wiley & Sons, New York, 1995.

[3] F. Black and M. Scholes. The pricing of options and corporate liabilities. *Journal of Political Economy* 81:637–654, 1973.

[4] Y.S. Chow and H. Teicher. *Probability Theory: Independence, Interchangeability, Martingales*. Springer-Verlag, New York, 2003.

[5] K.L. Chung. *A Course in Probability Theory*. Academic Press, New York, 2000.

[6] J.C. Cox, J.E. Ingersoll, and S.A. Ross. An intertemporal general equilibrium model of asset prices. *Econometrica* 53:363–384, 1985.

[7] D. Duffie. *Dynamic Asset Pricing Theory*. Princeton University Press, Princeton, NJ, 1996.

[8] D. Duffie and P. Glynn. Efficient Monte Carlo simulation for security prices. *Annals of Applied Probability* 5:897–905, 1995.

[9] P. Dupuis and H. Wang. Importance sampling, large deviations, and differential games. *Stochastics and Stochastics Reports* 76:481–508, 2004.

[10] P. Dupuis and H. Wang. Subsolutions of an Isaacs equation and efficient schemes for importance sampling. *Mathematics of Operations Research* 32:723–757, 2007.

[11] P. Glasserman. *Monte Carlo Methods in Financial Engineering*. Springer-Verlag, New York, 2004.

[12] P. Glasserman, P. Heidelberger, and P. Shahabuddin. Asymptotically optimal importance sampling and stratification for path-dependent options. *Mathematical Finance* 9:117–152, 1999.

[13] P. Glasserman and Y. Wang. Counter examples in importance sampling for large deviations probabilities. *Annals of Applied Probability* 7:731–746, 1997.

[14] G. Hardy, J.E. Littlewood, and G. Pólya. *Inequalities*. Cambridge University Press, Cambridge, UK, 1952.

[15] S.L. Heston. A closed-form solution for options with stochastic volatility with applications to bonds and currency options. *The Review of Financial Studies* 6:327–343, 1993.

[16] J. Hull. *Options, Futures, and Other Derivative Securities*. Prentice-Hall, New Jersey, 2000.

[17] S. Kou. A jump diffusion model for option pricing. *Management Sciences* 48:1086–1101, 2002.

[18] I. Karatzas and S.E. Shreve. *Brownian Motion and Stochastic Calculus*. Springer-Verlag, New York, 1999.

[19] I. Karatzas and S.E. Shreve. *Methods of Mathematical Finance*. Springer-Verlag, New York, 1998.

[20] P.E. Kloeden and E. Platen. *Numerical Solution of Stochastic Differential Equations*. Springer-Verlag, Berlin, 1992.

[21] H.J. Kushner and P. Dupuis. *Numerical Methods for Stochastic Control Problems in Continuous Time*. Springer-Verlag, New York, 1992.

[22] R.C. Merton. Option pricing when underlying stock returns are discontinuous. *Journal of Financial Economics* 3:125–144, 1976.

[23] R.C. Merton. *Continuous-Time Finance*. Blackwell Publishers, Cambridge, MA, 1994.

[24] S. Natenberg. *Option Volatility and Pricing*. McGraw-Hill, New York, 1994.

[25] B. Oksendal. *Stochastic Differential Equations: An Introduction with Applications*. Springer, Berlin, 1998.

[26] L.C.G. Rogers and D. Williams. *Diffusions, Markov Processes, and Martingales: Volume 1, Foundations.* Cambridge University Press, Cambridge, UK, 2001.

[27] L.C.G. Rogers and D. Williams. *Diffusions, Markov Processes, and Martingales: Volume 2, Itô Calculus.* Cambridge University Press, Cambridge, UK, 2001.

[28] R.Y. Rubinstein. Optimization of computer simulation models with rare events. *European Journal of Operations Research* 99:89–112, 1997.

[29] R.Y. Rubinstein and D.P. Kroese. *Simulation and the Monte Carlo Method.* John Wiley & Sons, New Jersey, 2008.

[30] R.U. Seydel. *Tools for Computational Finance.* Springer-Verlag, Berlin, 2009.

[31] D. Siegmund. Importance sampling in the Monte Carlo study of sequential tests. *Annals of Statistics* 4:673–684, 1976.

[32] Y.L. Tong. *The Multivariate Normal Distribution.* Springer-Verlag, New York, 1989.

[33] S.R.S. Varadhan. *Large Deviations and Applications.* CBMS-NSF Regional Conference Series in Mathematics, SIAM, Philadelphia, 1984.

[34] D.D. Wackerly, W. Mendenhall, and R.L. Scheaffer. *Mathematical Statistics with Applications.* Duxbury Press, USA, 2008.

Index

Made in the USA
Middletown, DE
24 January 2021